GENE CLONING

Art Center College of Design
Library
1700 Lida Street
Pasadena, Calif. 91103

DISCARD

ART CENTER COLLEGE OF DESIGN

3 3220 00209 4824

Art Center College of Design
Library
1700 Lida Street
Pasadena, Calif. 91103

GENE CLONING
an introduction

572.8633
B881
1998

Third edition

T.A. Brown

UMIST, Manchester, UK

Stanley Thornes (Publishers) Ltd

© 1986, 1990, 1995 T. A. Brown

The right of T. A. Brown to be identified as author of this work has been asserted by him in accordance with the Copyright, Designs and Patents Act 1988.

All rights reserved. No part of this publication may be reproduced or transmitted in any form or by any means, electronic or mechanical, including photocopying, recording or any information storage and retrieval system, without permission in writing from the publisher or under licence from the Copyright Licensing Agency Limited. Further details of such licences (for reprographic reproduction) may be obtained from the Copyright Licensing Agency Limited, of 90 Tottenham Court Road, London W1P 0LP.

First edition published by Chapman & Hall 1986
Second edition published by Chapman & Hall 1990
Third edition published by Chapman & Hall 1995

Reprinted in 1998 by:
Stanley Thornes (Publishers) Ltd
Ellenborough House
Wellington Street
CHELTENHAM
GL50 1YW
United Kingdom

98 99 00 01 02 / 10 9 8 7 6 5 4 3 2 1

A catalogue record for this book is available from the British Library

ISBN 0–7487–4070–8

Typeset by Hope Services (Abingdon) Ltd
Printed and bound in Italy by STIGE, Turin

Contents

Art Center College of Design
Library
1700 Lida Street
Pasadena, Calif. 91103

Art Center College of Design
Library
1700 Lida Street
Pasadena, Calif. 91103

Preface

The Third Edition of this book is longer and more detailed than either of its predecessors, but it retains the same basic philosophy. It is still unashamedly introductory and is still aimed at undergraduates and other individuals who have no previous experience of experiments with DNA. The last few years have seen a proliferation of cloning manuals and other 'hands-on' texts for gene cloners, but few of these address the needs of a student encountering the subject for the first time. I hope that this new Edition will continue to help these newcomers get started.

I have made revisions and updates at many places throughout the book, but major changes do not occur until the last few chapters. A major weakness of the Second Edition was the poor coverage that I had given to the polymerase chain reaction, my excuse being that the sudden rise of the technique occurred just as I had completed the manuscript. The Third Edition attempts to correct this failing with a new chapter devoted entirely to PCR. I appreciate that PCR is not really a component of gene cloning, so the title of the book is no longer appropriate, but 'Gene Cloning, the Polymerase Chain Reaction, and Related Techniques: An Introduction' seemed a bit long-winded. The 'Gene Cloning' of the title is now something of a generic term, and I apologize to any purists among the readership.

The other major rewrites are in Part Three where again I have extended the coverage, this time by bringing in a new chapter on the applications of gene cloning in agriculture, or to be precise, in plant genetic engineering. Here, and at a few other places in the book, I have attempted in a rather hesitant (and I fear inadequate) fashion to discuss some of the broader issues that arise from our ability to clone genes. These issues are now so prominent in the public perception that they must be addressed by all students of genetic engineering. So the real title of the book is 'Gene Cloning, the Polymerase Chain Reaction, Related Techniques and some of the Implications: An Introduction'.

As with the First and Second Editions, I would not have got very far with this book if my wife Keri had not been prepared to put up with several months of lonely evenings as I struggled with the word processor. Once again her unwavering encouragement has been the most important factor in completion of the task.

T. A. Brown
Manchester

Preface to the First Edition

This book is intended to introduce gene cloning and recombinant DNA technology to undergraduates who have no previous experience of the subject. As such, it assumes very little background knowledge on the part of the reader – just the fundamental details of DNA and genes that would be expected of an average sixth-former capable of a university entrance grade at A-level biology. I have tried to explain all the important concepts from first principles, to define all unfamiliar terms either in the text or in the glossary, to avoid the less-helpful jargon words, and to reinforce the text with as many figures as are commensurate with a book of reasonable price.

Although aimed specifically at first- and second-year undergraduates in biochemistry and related degree courses, I hope that this book will also prove useful to some experienced researchers. I have been struck over the last few years by the number of biologists, expert in other aspects of the science, who have realized that gene cloning may have a role in their own research projects. Possibly this text can act as a painless introduction to the complexities of recombinant DNA technology for those of my colleagues wishing to branch out into this new discipline.

I would like to make it clear that this book is not intended as competition for the two excellent gene cloning texts already on the market. I have considerable regard for the books by Drs Old and Primrose and by Professor Glover, but believe that both texts are aimed primarily at advanced undergraduates who have had some previous exposure to the subject. It is this 'previous exposure' that I aim to provide. My greatest satisfaction will come if this book is accepted as a primer for Old and Primrose or for Glover.

I underestimated the effort needed to produce such a book and must thank several people for their help. The publishers provided the initial push to get the project under way. I am indebted to Don Grierson at Nottingham University and Paul Sims at UMIST for reading the text and suggesting improvements; all errors and naïveties are, however, mine. Finally, my wife Keri typed most of the manuscript and came to my rescue on several occasions with the right word or turn of phrase. This would never have been finished without her encouragement.

T. A. Brown
Manchester

Preface to the Second Edition

It was only when I started writing the Second Edition to this book that I fully appreciated how far gene cloning has progressed since 1986. Being caught up in the day-to-day excitement of biological research it is sometimes difficult to stand back and take a considered view of everything that is going on. The pace with which new techniques have been developed and applied to recombinant DNA research is quite remarkable. Procedures which in 1986 were new and innovative are now *de rigueur* for any self-respecting research laboratory and many of the standard techniques have found their way into undergraduate practical classes. Students are now faced with a vast array of different procedures for cloning genes and an even more diverse set of techniques for studying them once they have been cloned.

In revising this book I have tried to keep rigidly to a self-imposed rule that I would not make the Second Edition any more advanced than the first. There are any number of advanced texts for students or research workers who need detailed information on individual techniques and approaches. In contrast there is still a surprising paucity of really introductory texts on gene cloning. The First Edition was unashamedly introductory and I hope that the Second Edition will be also.

Nevertheless, changes were needed and on the whole the Second Edition contains more information. I have resisted the temptation to make many additions to Part One, where the fundamentals of gene cloning are covered. A few new vectors are described, especially for cloning in eukaryotes, but on the whole the first seven chapters are very much as they were in the First Edition. Part Two has been redefined so it now concentrates more fully on techniques for studying cloned genes, in particular with a description of methods for analysing gene regulation. Recombinant DNA techniques in general have become more numerous since 1986 and an undergraduate is now expected to have a broader appreciation of how cloned genes are studied. In Part Three the main theme is still biotechnology, but the tremendous advances in this area have required more extensive rewriting. The use of eukaryotes for synthesis of recombinant protein is now standard procedure, and we have seen the first great contributions of gene cloning to the study of human disease. The applications of gene cloning really make up a different book to this one, but nonetheless in Part Three I have tried to give a flavour of what is going on.

A number of people have been kind enough to comment on the First Edition and make suggestions for this revision. Don Grierson and Paul Sims again provided important and sensible advice. I must also thank Stephen Oliver and Richard Walmsley for their comments on specific parts of the book. Once again my wife's patience and encouragement has been a major factor in getting a Second Edition done at all. Finally I would like to thank all the students who have used the First Edition for the mainly nice things they have said about it.

T. A. Brown
Manchester

Part One
The Basic Principles of Gene Cloning

Why gene cloning is important 1

Just over a century ago, Gregor Mendel formulated a set of rules to explain the inheritance of biological characteristics. The basic assumption of these rules is that each heritable property of an organism is controlled by a factor, called a **gene**, that is a physical particle present somewhere in the cell. The rediscovery of Mendel's laws in 1900 marks the birth of **genetics**, the science aimed at understanding what these genes are and exactly how they work.

1.1 THE EARLY DEVELOPMENT OF GENETICS

For the first 30 years of its life this new science grew at an astonishing rate. The idea that genes reside on **chromosomes** was proposed by W. Sutton in 1903, and received experimental backing from T. H. Morgan in 1910. Morgan and his colleagues then developed the techniques for **gene mapping**, and by 1922 had produced a comprehensive analysis of the relative positions of over 2000 genes on the four chromosomes of the fruit fly, *Drosophila melanogaster*.

Despite the brilliance of these classical genetic studies, there was no real understanding of the molecular nature of the gene until the 1940s. Indeed, it was not until the experiments of Avery, MacLeod and McCarty in 1944, and of Hershey and Chase in 1952, that anyone believed DNA to be the genetic material; up to then it was widely thought that genes were made of protein. The discovery of the role of DNA was a tremendous stimulus to genetic research, and many famous biologists (Delbrück, Chargaff, Crick and Monod were among the most influential) contributed to the second great age of genetics. In the 14 years between 1952 and 1966 the structure of DNA was elucidated, the genetic code cracked, and the processes of transcription and translation described.

1.2 THE ADVENT OF GENE CLONING

These years of activity and discovery were followed by a lull, a period of anticlimax when it seemed to some molecular biologists (as the new generation of geneticists styled themselves) that there was little of fundamental importance that was not understood. In truth there was a frustration that the experimental techniques of the late 1960s were not sophisticated enough to allow the gene to be studied in any greater detail.

Then, in the years 1971–1973 genetic research was thrown back into gear by what at the time was described as a revolution in experimental biology. A whole new methodology was developed, enabling previously impossible experiments to be planned and carried out, if not with ease, then at least with success. These methods, referred to as **recombinant DNA technology** or **genetic engineering**, and having at their core the process of **gene cloning**, sparked the third great age of genetics. Twenty-five years later we are still riding the rollercoaster set in motion by the gene cloning revolution, and there is no end to the excitement in sight.

1.3 WHAT IS GENE CLONING?

The basic steps in a gene cloning experiment are as follows (Figure 1.1).

1. A fragment of DNA, containing the gene to be cloned, is inserted into a circular DNA molecule called a **vector**, to produce a **chimaera** or **recombinant DNA molecule**.
2. The vector acts as a **vehicle** that transports the gene into a host cell, which is usually a bacterium, although other types of living cell can be used.
3. Within the host cell the vector multiplies, producing numerous identical copies not only of itself but also of the gene that it carries.
4. When the host cell divides, copies of the recombinant DNA molecule are passed to the progeny and further vector replication takes place.
5. After a large number of cell divisions, a colony, or **clone**, of identical host cells is produced. Each cell in the clone contains one or more copies of the recombinant DNA molecule; the gene carried by the recombinant molecule is now said to be cloned.

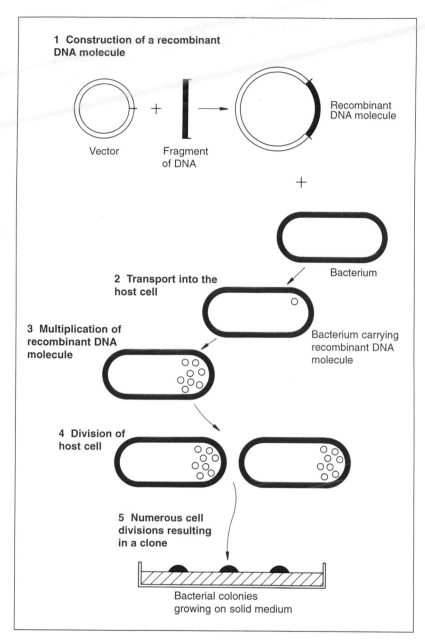

Figure 1.1 The basic steps in gene cloning.

1 **Construction of a recombinant DNA molecule**

Vector

Fragment of DNA

Recombinant DNA molecule

+

Bacterium

2 **Transport into the host cell**

3 **Multiplication of recombinant DNA molecule**

Bacterium carrying recombinant DNA molecule

4 **Division of host cell**

5 **Numerous cell divisions resulting in a clone**

Bacterial colonies growing on solid medium

1.4 GENE CLONING REQUIRES SPECIALIZED TOOLS AND TECHNIQUES

1.4.1 Vehicles

The central component of a gene cloning experiment is the vehicle, which transports the gene into the host cell and is responsible for its replication. To act as a cloning vehicle a DNA molecule must be capable of entering a host cell and, once inside, replicating to produce multiple copies of itself. Two naturally occurring types of DNA molecule satisfy these requirements:

1. **Plasmids**, which are small circles of DNA found in bacteria and some other organisms. Plasmids can replicate independently of the host cell chromosome.
2. **Virus chromosomes**, in particular the chromosomes of **bacteriophages**, which are viruses that specifically infect bacteria. During infection the bacteriophage DNA molecule is injected into the host cell where it undergoes replication.

Chapter 2 covers the basic features of plasmids and bacteriophage chromosomes, providing the necessary background for an understanding of how these molecules are used as cloning vehicles.

1.4.2 Techniques for handling DNA

Plasmids and bacteriophage DNA molecules display the basic properties required of potential cloning vehicles. But this potential would be wasted without experimental techniques for handling DNA molecules in the laboratory. The fundamental steps in gene cloning, as described on p. 4 and in Figure 1.1, require several manipulative skills (Table 1.1). First, pure samples of DNA must be available, both of the cloning vehicle and of the gene to be cloned. The methods used to purify DNA from living cells are outlined in Chapter 3.

Table 1.1 Basic skills needed to carry out a simple gene cloning experiment

1. Preparation of pure samples of DNA	(Chapter 3)
2. Cutting DNA molecules	(Chapter 4, pp. 60–68)
3. Analysis of DNA fragment sizes	(Chapter 4, pp. 68–76)
4. Joining DNA molecules together	(Chapter 4, pp. 77–85)
5. Introduction of DNA into host cells	(Chapter 5, pp. 87–93)
6. Identification of cells that contain recombinant DNA molecules	(Chapter 5, pp. 94–104, and Chapter 8)

Having prepared samples of DNA, construction of a recombinant DNA molecule requires that the vector be cut at a specific point and then repaired in such a way that the gene is inserted into the vehicle. The ability to manipulate DNA in this way is an offshoot of basic research into DNA synthesis and modification within living cells. The discovery of enzymes that can cut or join DNA molecules in the cell has led to the purification of **restriction endonucleases** and **ligases**, which are now used to construct recombinant DNA molecules in the test-tube. The properties of these enzymes, and the way they are used in gene cloning experiments, are described in Chapter 4.

Once a recombinant DNA molecule has been constructed, it must be introduced into the host cell so that replication can take place. Transport into the host cell makes use of natural processes for uptake of plasmid and viral DNA molecules. These processes, and the ways they are utilized in gene cloning, are described in Chapter 5.

1.4.3 The diversity of cloning vectors

Although gene cloning is relatively new, it has nevertheless developed into a very sophisticated technology. Today a wide variety of different cloning vectors are available. Almost all of these are derived from naturally occurring plasmids or viruses, but most have been modified in various ways so that each one is suited for a particular type of cloning experiment. In Chapters 6 and 7 the most important types of vector are described, and their uses examined.

1.5 WHY GENE CLONING IS SO IMPORTANT

As you can see from Figure 1.1, gene cloning is a relatively straightforward procedure. Why then has it assumed such importance in biology? The answer is largely because cloning can provide a pure sample of an individual gene, separated from all the other genes that it normally shares the cell with.

To understand exactly how this works, consider a gene cloning experiment drawn in a slightly different way (Figure 1.2). In this example the DNA fragment to be cloned is one member of a mixture of many different fragments, each carrying a different gene or part of a gene. This mixture could indeed be the entire genetic complement of an organism, a human for instance. All these fragments will become inserted into different vector molecules to produce a family of recombinant DNA molecules, one of which carries the gene of interest. Usually only one recombinant DNA molecule will be transported into any single host cell, so that although the final set of clones may contain many different recombinant DNA molecules, each individual clone contains multiple copies of just one molecule. The gene is now separated away from all the other genes in the original mixture, and its specific features can be studied in detail.

In practice, the key to the success or failure of a cloning experiment is the ability to identify the particular clone of interest from the many different ones that are obtained. If we consider the

Figure 1.2 Cloning allows individual fragments of DNA to be purified.

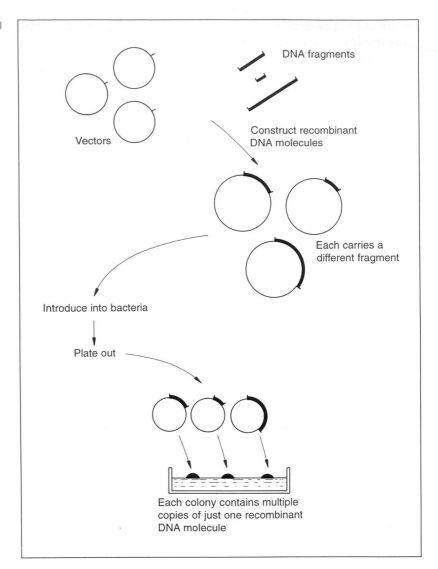

genome of the bacterium *Escherichia coli*, which contains something in the region of 4000 different genes, we might at first despair of being able to find just one gene among all the possible clones (Figure 1.3). The problem becomes even more overwhelming when we remember that bacteria are relatively simple organisms and that the human genome contains about 20 times as many genes. However, as explained in Chapter 8, a variety of different strategies can be used to ensure that the correct gene can be obtained at the end of the cloning experiment. Some of these strategies involve modifications to the basic cloning procedure, so that only cells containing the desired recombinant DNA molecule can divide and the clone of interest is automatically **selected**. Other methods involve techniques that enable the desired clone to be identified from a mixture of lots of different clones.

Once a gene has been cloned there is almost no limit to the information that can be obtained about the structure and expression of that gene. The availability of cloned material has stimulated the development of analytical methods for studying genes, with new techniques being introduced all the time. Methods for studying the structure and expression of a cloned gene are described in Chapters 9 and 10 respectively.

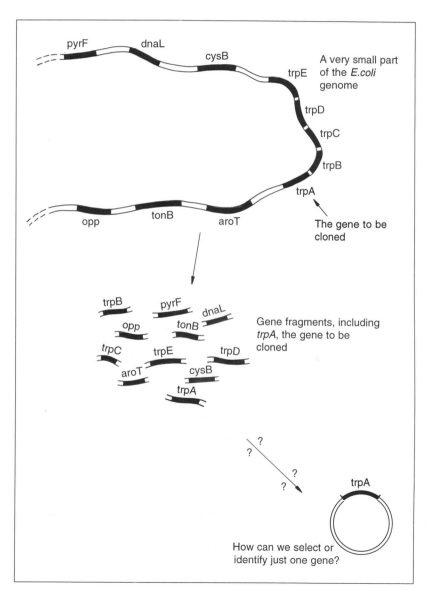

Figure 1.3 The problem of selection.

1.6 WHY THE POLYMERASE CHAIN REACTION IS ALSO IMPORTANT

Until the late 1980s cloning was the only means of obtaining a pure sample of an individual gene. This is no longer the case as the exquisitely simple but extraordinarily powerful **polymerase chain reaction (PCR)** provides today's biologists with a second approach to gene isolation. In a PCR experiment a single segment of a DNA molecule is copied many times, resulting in an amplified DNA fragment (Figure 1.4). The experiment is designed so that the segment of DNA that is amplified is one that carries a gene of interest. The result is therefore the same as with cloning: a pure sample of a single gene is obtained. PCR is a rapid technique, much less complicated than gene cloning, and can amplify millions of copies of a gene from just one starting molecule. This tremendous sensitivity means that PCR can be used to isolate genes from single cells, or from material such as dried bloodstains or even the bones of long-dead humans. The polymerase chain reaction and its many applications are covered in Chapter 11.

Figure 1.4 The polymerase chain reaction enables an individual gene to be amplified.

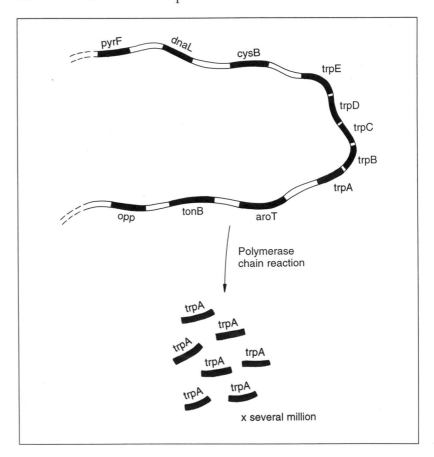

1.7 THE IMPACT OF RECOMBINANT DNA TECHNIQUES ON RESEARCH AND BIOTECHNOLOGY

There are very few areas of biological research that have not been touched by gene cloning, PCR, and the recombinant DNA techniques that these procedures have made possible. In industry, for example, the ability to clone genes has led to far-reaching advances in **biotechnology**. For many years microorganisms have been used as living factories for the production of useful compounds. Examples are provided by antibiotics, such as penicillin, which is synthesized by a fungus called *Penicillium*, and streptomycin, produced by the bacterium *Streptomyces griseus*. Gene cloning has revolutionized biotechnology, most notably in providing a way in which mammalian proteins can be produced in bacterial cells. A remarkable property of a cloned gene is that it can often be made to function in an organism totally unrelated to that in which it is normally found. For example, an animal gene can be transferred by cloning into a bacterium and then induced by some careful modifications to carry on working as though nothing had happened (Figure 1.5). The implications are enormous. Genes controlling the

Figure 1.5 A possible scheme for the production of an animal protein by a bacterium.

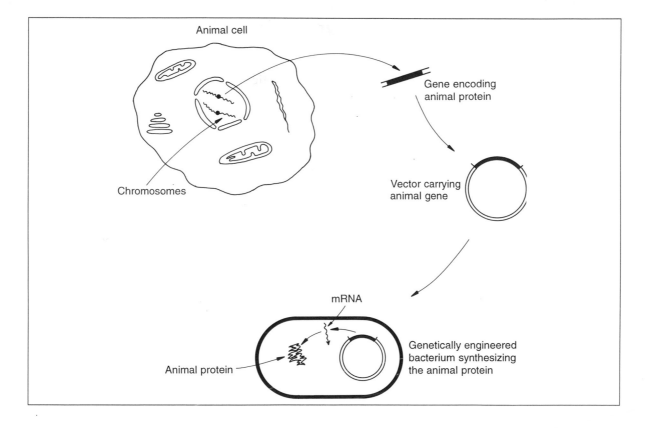

synthesis of important pharmaceuticals, such as drugs and hormones, can be taken from the organism in which they occur naturally, but from which they may be costly and difficult to prepare, and placed in a bacterium or other type of organism, from which the product can be recovered conveniently and in large quantities. A number of successes have been notched up by biotechnologists, with recombinant insulin being the most noteworthy achievement so far. Production of protein by cloned genes is described in Chapter 12.

Medical and agricultural research have also received important boosts from gene cloning. New types of vaccines, providing protection against diseases for which vaccination was previously impossible, have been developed thanks to the ability to clone genes. Many inherited diseases can now be diagnosed in an unborn child, and recent research has led to the hope that cystic fibrosis, breast cancer and other heart-rending diseases will soon be treatable. In agriculture, equally important problems are being addressed through the development of genetically engineered crops able to withstand the ravages of insects. The ways in which gene cloning is being applied in these areas of research are described in the final two chapters of this book.

FURTHER READING

Cherfas, J. (1982) *Man Made Life*, Blackwell, Oxford, UK – a history of the early years of genetic engineering.

Felsenfeld, G. (1985) DNA, in *The Molecules of Life*, Readings from *Scientific American*, Freeman, New York, USA; also Felsenfeld, G. (1985) DNA. *Scientific American*, **253** (Oct): 44–53 – a brief review of the important features of DNA.

Brown, T. A. (1992) *Genetics: A Molecular Approach*, 2nd edn, Chapman & Hall, London, UK – an introduction to genetics and molecular biology.

Kendrew, A. (ed.) (1994) *The Encyclopedia of Molecular Biology*. Blackwell Science, Oxford, UK – an outstanding compendium of everything you need to know about molecular biology.

Beebee, T. and Burke, J. (1992) *Gene Structure and Transcription*, 2nd edn, IRL Press at Oxford University Press, Oxford, UK – details of genes and gene expression.

Vehicles: plasmids and bacteriophages

2

A DNA molecule needs to display several features to be able to act as a vehicle for gene cloning. Most important, it must be able to replicate within the host cell, so that numerous copies of the recombinant DNA molecule can be produced and passed to the daughter cells. A cloning vehicle also needs to be relatively small, ideally less than 10 kilobases (kb) in size, as large molecules tend to break down during purification, and are also more difficult to manipulate. Two kinds of DNA molecule that satisfy these criteria can be found in bacterial cells: plasmids and bacteriophage chromosomes. Although plasmids are frequently employed as cloning vehicles, two of the most important types of vector in use today are derived from bacteriophages.

2.1 PLASMIDS

2.1.1 Basic features of plasmids

Plasmids are circular molecules of DNA that lead an independent existence in the bacterial cell (Figure 2.1). Plasmids almost always carry one or more genes, and often these genes are responsible for

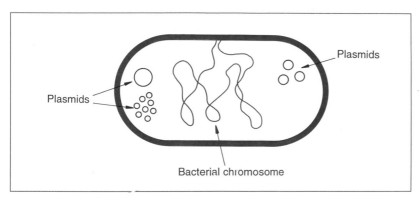

Figure 2.1 Plasmids: independent genetic elements found in bacterial cells.

a useful characteristic displayed by the host bacterium. For example, the ability to survive in normally toxic concentrations of antibiotics such as chloramphenicol or ampicillin is often due to the presence in the bacterium of a plasmid carrying antibiotic-resistance genes. In the laboratory antibiotic resistance is often used as a **selectable marker** to ensure that bacteria in a culture contain a particular plasmid (Figure 2.2).

All plasmids possess at least one DNA sequence that can act as an **origin of replication**, so they are able to multiply within the cell quite independently of the main bacterial chromosome (Figure

Figure 2.2 The use of antibiotic resistance as a selectable marker for a plasmid. RP4 (top) carries genes for resistance to ampicillin, tetracycline and kanamycin. Only those *E. coli* cells that contain RP4 (or a related plasmid) are able to survive and grow in a medium that contains toxic amounts of one or more of these antibiotics.

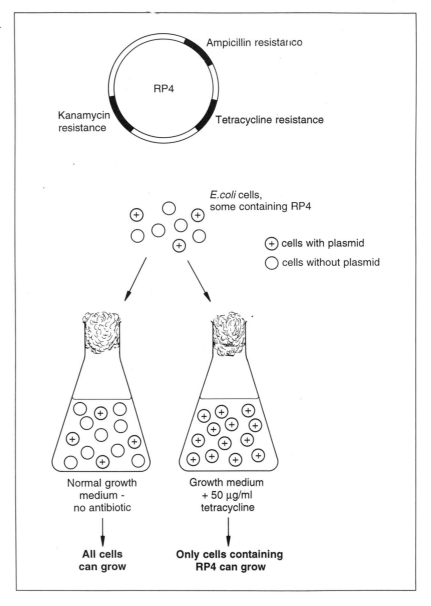

2.3(a)). The smaller plasmids make use of the host cell's own DNA replicative enzymes in order to make copies of themselves, whereas some of the larger ones carry genes that code for special enzymes that are specific for plasmid replication.

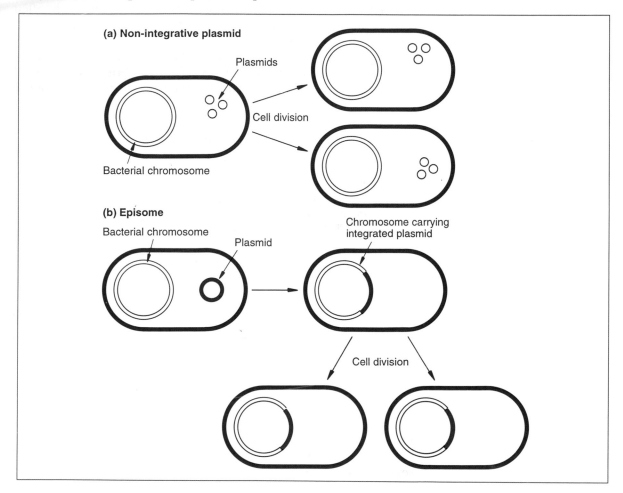

A few types of plasmid are also able to replicate by inserting themselves into the bacterial chromosome (Figure 2.3(b)). These integrative plasmids or **episomes** may be stably maintained in this form through numerous cell divisions, but will at some stage exist as independent elements. Integration is also an important feature of some bacteriophage chromosomes and will be described in more detail when these are considered (p. 20).

Figure 2.3 Replication strategies for (a) a non-integrative plasmid, and (b) an episome.

2.1.2 Size and copy number

These two features of plasmids are particularly important as far as cloning is concerned. We have already mentioned the relevance of plasmid size and stated that less than 10 kb is desirable for a cloning vehicle. Plasmids range from about 1.0 kb for the smallest to over 250 kb for the largest plasmids (Table 2.1), so only a few will be useful for cloning purposes. However, as described in Chapter 7, larger plasmids may be adapted for cloning under some circumstances.

Table 2.1 Sizes of representative plasmids

Plasmid	Size		Organism
	Nucleotide length (kb)	Molecular wt (MDa)	
pUC8	2.1	1.8	E. coli
ColEI	6.4	4.2	E. coli
RP4	54	36	Pseudomonas + others
F	95	63	E. coli
TOL	117	78	Pseudomonas putida
pTiAch5	213	142	Agrobacterium tumefaciens

The **copy number** refers to the number of molecules of an individual plasmid that are normally found in a single bacterial cell. The factors that control copy number are not well understood, but each plasmid has a characteristic value that may be as low as one (especially for the large molecules) or as many as 50 or more. Generally speaking, a useful cloning vehicle needs to be present in the cell in multiple copies so that large quantities of the recombinant DNA molecule can be obtained.

2.1.3 Conjugation and compatibility

Plasmids fall into two groups: conjugative and non-conjugative. Conjugative plasmids are characterized by the ability to promote sexual **conjugation** between bacterial cells (Figure 2.4), a process that can result in a conjugative plasmid spreading from one cell to all the other cells in a bacterial culture. Conjugation and plasmid transfer are controlled by a set of transfer or *tra* genes, which are present on conjugative plasmids but absent from the non-conjugative type. However, a non-conjugative plasmid may, under some circumstances, be cotransferred along with a conjugative plasmid when both are present in the same cell.

Several different kinds of plasmid may be found in a single cell, including more than one different conjugative plasmid at any one time. In fact, cells of *E. coli* have been known to contain up to seven different plasmids at once. To be able to coexist in the same cell, different plasmids must be **compatible**. If two plasmids are incompatible then one or the other will be quite rapidly lost from the cell.

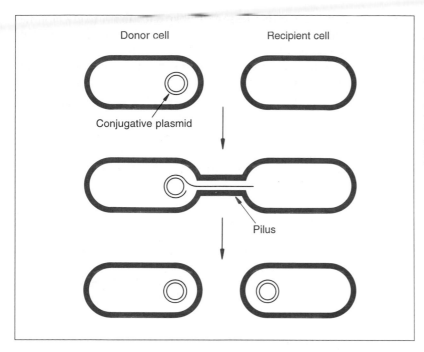

Figure 2.4 Plasmid transfer by conjugation between bacterial cells. The donor and recipient cells attach to each other by a pilus, a hollow appendage present on the surface of the donor cell. A copy of the plasmid is then passed to the recipient cell. Transfer is thought to occur through the pilus, as drawn in the Figure, but this has not been proven and transfer by some other means (e.g. directly across the bacterial cell walls) remains a possibility.

Different types of plasmid can therefore be assigned to different **incompatibility groups** on the basis of whether or not they can coexist, and plasmids from a single incompatibility group are often related to each other in various ways. The basis of incompatibility is not well understood, but events during plasmid replication are thought to underlie the phenomenon.

2.1.4 Plasmid classification

The most useful classification of naturally occurring plasmids is based on the main characteristic coded by the plasmid genes. The five main types of plasmid according to this classification are as follows.

1. **Fertility** or **'F' plasmids** carry only *tra* genes and have no characteristic beyond the ability to promote conjugal transfer of plasmids; e.g. F plasmid of *E. coli*.
2. **Resistance** or **'R' plasmids** carry genes conferring on the host bacterium resistance to one or more antibacterial agents, such as chloramphenicol, ampicillin and mercury. R plasmids are very important in clinical microbiology as their spread through natural populations can have profound consequences in the treatment of bacterial infections; e.g. RP4, commonly found in *Pseudomonas*, but also occurring in many other bacteria.
3. **Col plasmids** code for colicins – proteins that kill other bacteria; e.g. ColE1 of *E. coli*.

4. **Degradative plasmids** allow the host bacterium to metabolize unusual molecules such as toluene and salicylic acid; e.g. TOL of *Pseudomonas putida*.

5. **Virulence plasmids** confer pathogenicity on the host bacterium; e.g. **Ti plasmids** of *Agrobacterium tumefaciens*, which induce crown gall disease on dicotyledonous plants.

2.1.5 Plasmids in organisms other than bacteria

Although plasmids are widespread in bacteria they are by no means so common in other organisms. The best characterized eukaryotic plasmid is the **2 μm circle** that occurs in many strains of the yeast *Saccharomyces cerevisiae*. The discovery of the 2 μm plasmid was very fortuitous as it has allowed the construction of vectors for cloning genes with this very important industrial organism as the host (p. 133). However, the search for plasmids in other eukaryotes (e.g. filamentous fungi, plants and animals) has proved disappointing, and it is suspected that many higher organisms simply do not harbour plasmids within their cells.

2.2 BACTERIOPHAGES

2.2.1 Basic features of bacteriophages

Bacteriophages, or phages as they are commonly known, are viruses that specifically infect bacteria. Like all viruses, phages are very simple in structure, consisting merely of a DNA (or occasionally RNA) molecule carrying a number of genes, including several for replication of the phage, surrounded by a protective coat or **capsid** made up of protein molecules (Figure 2.5).

Figure 2.5 Schematic representations of the two main types of phage structure. (a) Head-and-tail (e.g. λ). (b) Filamentous (e.g. M13).

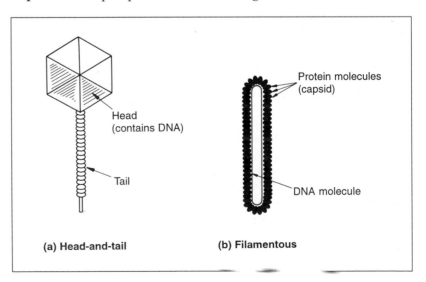

Head
(contains DNA)

Tail

Protein molecules
(capsid)

DNA molecule

(a) Head-and-tail   **(b) Filamentous**

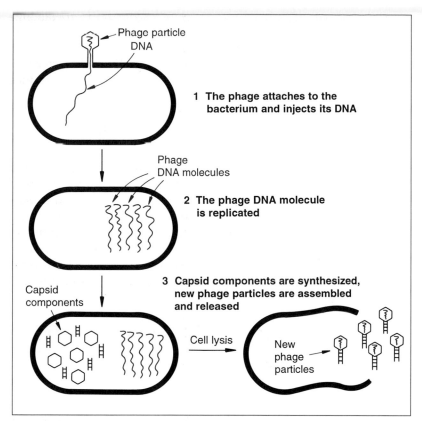

Phage particle
DNA

1 The phage attaches to the
   bacterium and injects its DNA

Phage
DNA molecules

2 The phage DNA molecule
   is replicated

Capsid
components

3 Capsid components are synthesized,
   new phage particles are assembled
   and released

Cell lysis

New
phage
particles

Figure 2.6 The general pattern of infection of a bacterial cell by a bacteriophage.

The general pattern of infection, which is the same for all types of phage, is a three-step process (Figure 2.6).

1. The phage particle attaches to the outside of the bacterium and injects its DNA chromosome into the cell.
2. The phage DNA molecule is replicated, usually by specific phage enzymes coded by genes on the phage chromosome.
3. Other phage genes direct synthesis of the protein components of the capsid, and new phage particles are assembled and released from the bacterium.

With some phage types the entire infection cycle is completed very quickly, possibly in less than 20 minutes. This type of rapid infection is called a **lytic cycle**, as release of the new phage particles is associated with lysis of the bacterial cell. The characteristic feature of a lytic infection cycle is that phage DNA replication is immediately followed by synthesis of capsid proteins, and the phage DNA molecule is never maintained in a stable condition in the host cell.

2.2.2 Lysogenic phages

In contrast to a lytic cycle, **lysogenic** infection is characterized by retention of the phage DNA molecule in the host bacterium, possibly for many thousands of cell divisions. With many lysogenic phages the phage DNA is inserted into the bacterial genome, in a manner similar to episomal insertion (see Figure 2.3(b)). The integrated form of the phage DNA (called the **prophage**) is quiescent, and a bacterium (referred to as a **lysogen**) which carries a prophage is usually physiologically indistinguishable from an uninfected cell. However, the prophage is eventually released from the host genome and the phage reverts to the lytic mode and lyses the cell. The infection cycle of λ, a typical lysogenic phage of this type, is shown in Figure 2.7.

A limited number of lysogenic phages follow a rather different infection cycle. When **M13**, or a related phage, infects *E. coli*, new phage particles are continuously assembled and released from the cell. The M13 DNA is not integrated into the bacterial genome and does not become quiescent. With these phages, cell lysis never occurs, and the infected bacterium can continue to grow and divide, albeit at a slower rate than uninfected cells. Figure 2.8 shows the M13 infection cycle.

Although there are many different varieties of bacteriophage, only λ and M13 have found any real role as cloning vectors. The properties of these two phages will now be considered in more detail.

(a) Gene organization in the λ DNA molecule   λ is a typical example of a head-and-tail phage (Figure 2.5(a)). The DNA is contained in the polyhedral head structure and the tail serves to attach the phage to the bacterial surface and to inject the DNA into the cell (Figure 2.7).

The λ DNA molecule is 49 kb in size and has been intensively studied by the techniques of gene mapping and **DNA sequencing**. As a result the positions and identities of most of the genes on the λ DNA molecule are known (Figure 2.9). A feature of the λ genetic map is that genes related in terms of function are clustered together on the genome. For example, all of the genes coding for components of the capsid are grouped together in the left-hand third of the molecule, and genes controlling integration of the prophage into the host genome are clustered in the middle of the molecule. Clustering of related genes is profoundly important for controlling expression of the λ genome, as it allows genes to be switched on and off as a group rather than individually. Clustering is also important in the construction of λ-based cloning vectors (described in Chapter 6).

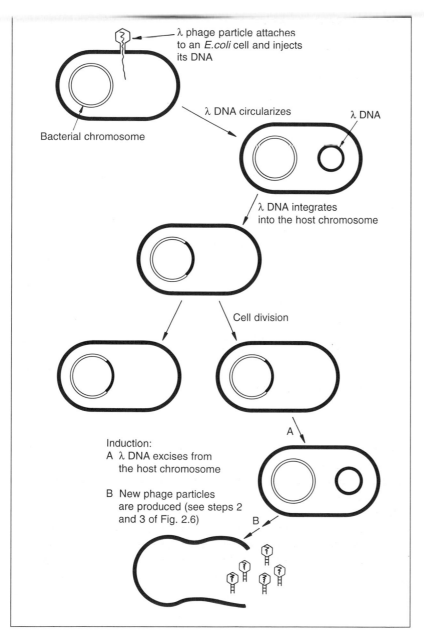

Figure 2.7 The lysogenic infection cycle of bacteriophage λ.

(Labels within figure:)

λ phage particle attaches to an *E.coli* cell and injects its DNA

Bacterial chromosome

λ DNA circularizes

λ DNA

λ DNA integrates into the host chromosome

Cell division

Induction:

A λ DNA excises from the host chromosome

B New phage particles are produced (see steps 2 and 3 of Fig. 2.6)

A

B

(b) The linear and circular forms of λ DNA   A second feature of λ that turns out to be of importance in the construction of cloning vectors is the conformation of the DNA molecule. The molecule shown in Figure 2.9 is linear, with two free ends, and represents the DNA present in the phage head structure. This linear molecule consists of two **complementary** strands of DNA, base-paired according to the **Watson–Crick rules** (that is,

Figure 2.8 The infection cycle of bacteriophage M13.

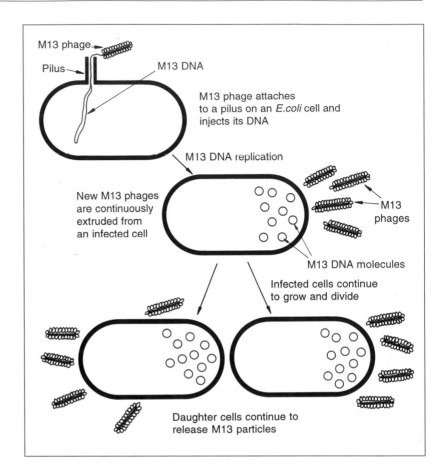

double-stranded DNA). However, at either end of the molecule is a short 12-nucleotide stretch, in which the DNA is single-stranded (Figure 2.10(a)). The two single strands are complementary, and so can base-pair with one another to form a circular, completely double-stranded molecule (Figure 2.10(b)).

Complementary single strands are often referred to as **'sticky'**

Figure 2.9 The λ genetic map, showing the positions of the important genes and the functions of the gene clusters.

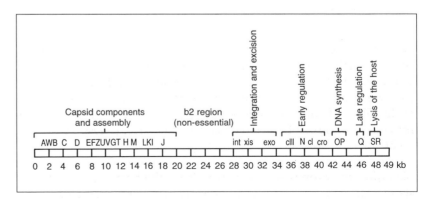

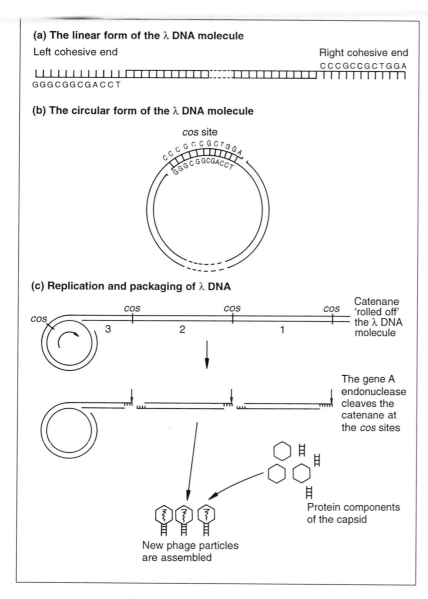

(a) The linear form of the λ DNA molecule

Left cohesive end

Right cohesive end
CCCGCCGCTGGA

GGCGGCGACCT

(b) The circular form of the λ DNA molecule

cos site

CCCGCCGCTGGA
GGGCGGCGACCT

(c) Replication and packaging of λ DNA

cos   cos   cos

cos

3   2   1

Catenane
'rolled off'
the λ DNA
molecule

The gene A
endonuclease
cleaves the
catenane at
the cos sites

Protein components
of the capsid

New phage particles
are assembled

Figure 2.10 The linear and circular forms of λ DNA. (a) The linear form showing the left and right cohesive ends. (b) Base-pairing between the cohesive ends results in the circular form of the molecule. (c) Rolling circle replication produces a catenane of new linear λ DNA molecules, which are individually packaged into phage heads as new λ particles are assembled.

ends or 'cohesive' ends, because base-pairing between them can 'stick' together the two ends of a DNA molecule (or the ends of two different DNA molecules). The λ cohesive ends are called the *cos* **sites** and they play two distinct roles during the λ infection cycle. First of all, they allow the linear DNA molecule that is injected into the cell to be circularized, which is a necessary prerequisite for insertion into the bacterial genome (Figure 2.7).

The second role of the *cos* sites is rather different, and comes into play after the prophage has excised from the host genome. At this stage a large number of new λ DNA molecules are produced by the rolling circle mechanism of replication (Figure 2.10(c)), in which a continuous DNA strand is 'rolled off' of the template molecule. The result is a catenane consisting of a series of linear λ genomes joined together at the *cos* sites. The role of the *cos* sites is now to act as recognition sequences for an **endonuclease** which cleaves the catenane at the *cos* sites producing individual λ genomes. This endonuclease (which is the product of gene A on the λ DNA molecule) creates the single-stranded sticky ends, and also acts in conjunction with other proteins to package each λ genome into a phage head structure.

As we shall see in Chapter 6, the cleavage and packaging processes recognize just the *cos* sites and the DNA sequences to either side of them. Changing the structure of the internal regions of the λ genome, for example by inserting new genes, has no effect on these events so long as the overall length of the λ genome is not altered too greatly.

(c) M13 – a filamentous phage M13 is an example of a filamentous phage (Figure 2.5(b)) and is completely different in structure from λ. Furthermore, the M13 DNA molecule is much smaller than the λ genome, being only 6407 nucleotides in length. It is circular, and is unusual in that it consists entirely of single-stranded DNA.

The smaller size of the M13 DNA molecule means that it has room for fewer genes than the λ genome. This is possible because the M13 capsid is constructed from multiple copies of just three proteins (requiring only three genes), whereas synthesis of the λ head-and-tail structure involves over 15 different proteins. In addition, M13 follows a simpler infection cycle than λ and does not need genes for insertion into the host genome.

Injection of an M13 DNA molecule into an *E. coli* cell occurs via the **pilus**, the structure that connects two cells during sexual conjugation (see Figure 2.4). Once inside the cell the single-stranded molecule acts as the template for synthesis of a complementary strand, resulting in normal double-stranded DNA (Figure 2.11(a)). This molecule is not inserted into the bacterial genome, but instead replicates until over 100 copies are present in the cell (Figure 2.11(b)). When the bacterium divides, each daughter receives copies of the phage genome, which continues to replicate, thereby maintaining its overall numbers per cell. As shown in Figure 2.11(c), new phage particles are continuously assembled and released, about 1000 new phages being produced during each generation of an infected cell.

(d) The attraction of M13 as a cloning vehicle Several features of M13 make this phage attractive as the basis for a cloning vehicle. The genome is less than 10 kb in size, well within the range

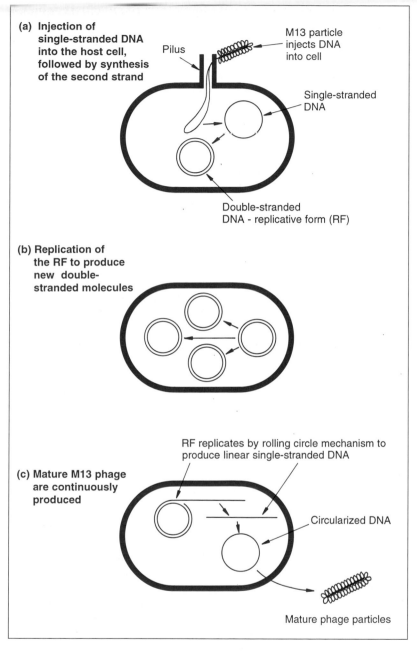

(a) Injection of single-stranded DNA into the host cell, followed by synthesis of the second strand

Pilus

M13 particle injects DNA into cell

Single-stranded DNA

Double-stranded DNA - replicative form (RF)

(b) Replication of the RF to produce new double-stranded molecules

RF replicates by rolling circle mechanism to produce linear single-stranded DNA

(c) Mature M13 phage are continuously produced

Circularized DNA

Mature phage particles

Figure 2.11 The M13 infection cycle showing the different types of DNA replication that occur. (a) After infection the single-stranded M13 DNA molecule is converted into the double-stranded replicative form (RF). (b) The RF replicates to produce multiple copies of itself. (c) Single-stranded molecules are synthesized by rolling circle replication and used in the assembly of new M13 particles.

that we stated was desirable for a potential vector. In addition, the double-stranded **replicative form** of the M13 genome behaves very much like a plasmid and can be treated as such for experimental purposes. It is easily prepared from a culture of infected *E. coli* cells (p. 36) and can be reintroduced by **transfection** (p. 99).

Most importantly, genes cloned with an M13-based vector can

be obtained in the form of single-stranded DNA. Single-stranded versions of cloned genes are useful for several techniques, notably DNA sequencing and *in vitro* mutagenesis (pp. 193 and 223). Using an M13 vector is an easy and reliable way of obtaining single-stranded DNA for this type of work.

2.2.3 Viruses as cloning vehicles for other organisms

Most living organisms are infected by viruses and it is not surprising that great interest has been shown in the possibility that viruses might be used as cloning vehicles for higher organisms. This is especially important when it is remembered that plasmids are not commonly found in organisms other than bacteria and yeast (p. 18).

In fact viruses have considerable potential as cloning vehicles for animal cells. Mammalian viruses such as **simian virus 40 (SV40)** and **adenoviruses**, and the insect **baculoviruses**, are the ones that have received most attention so far, but others are also being studied. These are discussed more fully in Chapter 7.

FURTHER READING

Dale, J. W. (1994) *Molecular Genetics of Bacteria*, 2nd edn, Wiley, Chichester, UK – provides a detailed description of plasmids and bacteriophages.

Purification of DNA from living cells

3

The genetic engineer will, at different times, need to prepare at least three distinct kinds of DNA. Firstly, **total cell DNA** will often be required as a source of material from which to obtain genes to be cloned. Total cell DNA may be DNA from a culture of bacteria, from a plant, from animal cells, or from any other type of organism that is being studied.

The second type of DNA that will be required is pure plasmid DNA. Preparation of plasmid DNA from a culture of bacteria follows the same basic steps as purification of total cell DNA, with the crucial difference that at some stage the plasmid DNA must be separated from the main bulk of chromosomal DNA also present in the cell.

Finally, phage DNA will be needed if a phage cloning vehicle is to be used. Phage DNA is generally prepared from bacteriophage particles rather than from infected cells, so there is no problem with contaminating bacterial DNA. However, special techniques are needed to remove the phage capsid. An exception is the double-stranded replicative form of M13 which is prepared from *E. coli* cells just as though it were a bacterial plasmid.

3.1 PREPARATION OF TOTAL CELL DNA

The fundamentals of DNA preparation are most easily understood by first considering the simplest type of DNA purification procedure, that where the entire DNA complement of a bacterial cell is required. The modifications needed for plasmid and phage DNA preparation can then be described later.

The procedure for total DNA preparation from a culture of bacterial cells can be divided into four stages (Figure 3.1).

1. A culture of bacteria is grown and then **harvested**.
2. The cells are broken open to release their contents.

Figure 3.1 The basic steps in preparation of total cell DNA from a culture of bacteria.

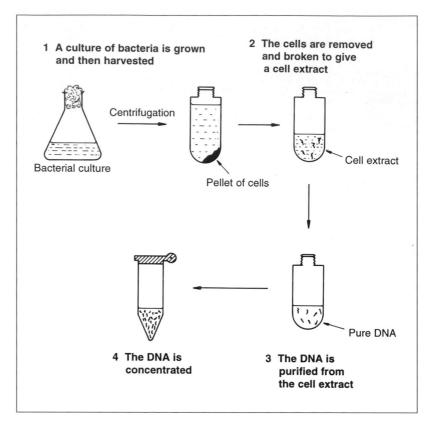

1 **A culture of bacteria is grown and then harvested**

2 **The cells are removed and broken to give a cell extract**

Centrifugation

Bacterial culture

Pellet of cells

Cell extract

Pure DNA

4 **The DNA is concentrated**

3 **The DNA is purified from the cell extract**

3. This **cell extract** is treated to remove all components except the DNA.
4. The resulting DNA solution is concentrated.

3.1.1 Growing and harvesting a bacterial culture

Most bacteria can be grown without too much difficulty in a liquid medium (**broth culture**). The culture medium must provide a balanced mixture of the essential nutrients at concentrations that will allow the bacteria to grow and divide efficiently. Two typical growth media are detailed in Table 3.1. M9 is an example of a **defined medium** in which all the components are known. This medium contains a mixture of inorganic nutrients to provide essential elements such as nitrogen, magnesium and calcium, as well as glucose to supply carbon and energy. In practice, additional growth factors, such as trace elements and vitamins, must be added to M9 before it will support bacterial growth. Precisely what supplements are needed depends on the species concerned.

The second medium described in Table 3.1 is rather different. LB is a complex or **undefined medium**, meaning that the precise iden-

Table 3.1   The composition of two typical media for the growth of bacterial cultures

Component	g/l
1. M9 medium	
Na_2HPO_4	6.0
KH_2PO_4	3.0
NaCl	0.5
NH_4Cl	1.0
$MgSO_4$	0.5
glucose	2.0
$CaCl_2$	0.015
2. LB (Luria–Bertani) medium	
tryptone	10
yeast extract	5
NaCl	10

tity and quantity of its components are not known. This is because two of the ingredients, tryptone and yeast extract, are complicated mixtures of unknown chemical compounds. Tryptone in fact supplies amino acids and small peptides, while yeast extract (a dried preparation of partially digested yeast cells) provides the nitrogen requirements, along with sugars and inorganic and organic nutrients. Complex media such as LB need no further supplementation and support the growth of a wide range of bacterial species.

Defined media must be used when the bacterial culture has to be grown under precisely controlled conditions. However, this is not necessary when the culture is being grown simply as a source of DNA, and under these circumstances a complex medium is appropriate. In LB medium at 37°C, aerated by shaking at 150–250 r.p.m. on a rotary platform, *E. coli* cells divide once every 20 minutes or so until the culture reaches a maximum density of about $2–3 \times 10^9$ cells/ml. The growth of the culture can be monitored by reading the optical density (OD) at 600 nm (Figure 3.2), at which wavelength one OD unit corresponds to about 0.8×10^9 cells/ml.

In order to prepare a cell extract, the bacteria must be obtained in as small a volume as possible. Harvesting is therefore performed by spinning the culture in a centrifuge (Figure 3.3). Fairly low centrifugation speeds will pellet the bacteria at the bottom of the centrifuge tube, allowing the culture medium to be poured off. Bacteria from a 1000 ml culture at maximum cell density can then be resuspended into a volume of 10 ml or less.

Figure 3.2 Estimation of bacterial cell number by measurement of optical density. (a) A sample of the culture is placed in a glass cuvette and light with a wavelength of 600 nm shone through. The amount of light that passes through the culture is measured and the optical density (also called the absorbance) calculated as 1 OD unit = $-\log_{10}$ (intensity of transmitted light)/(intensity of incident light). The operation is performed with a spectrophotometer. (b) The cell number corresponding to the OD reading is calculated from a calibration curve. This curve is plotted from the OD values of a series of cultures of known cell density. For *E. coli* 1 OD unit $\approx 0.8 \times 10^9$ cells/ml.

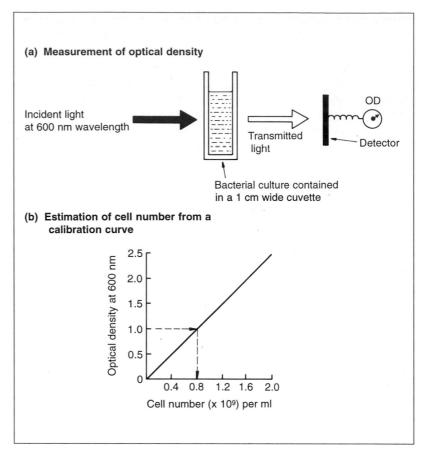

3.1.2 Preparation of a cell extract

The bacterial cell is enclosed in a cytoplasmic membrane and surrounded by a rigid cell wall. With some species, including *E. coli*, the cell wall may itself be enveloped by a second, outer membrane. All of these barriers have to be disrupted to release the cell components.

Techniques for breaking open bacterial cells can be divided into physical methods, in which the cells are disrupted by mechanical forces, and chemical methods, where cell lysis is brought about by exposure to chemical agents that affect the integrity of the cell barriers. Chemical methods are most commonly used with bacterial cells when the object is DNA preparation.

Chemical lysis generally involves one agent attacking the cell wall and another disrupting the cell membrane (Figure 3.4(a)). The chemicals that are used depend on the species of bacterium involved, but with *E. coli* and related organisms, weakening of the cell wall is usually brought about by **lysozyme**, ethylenediamine tetraacetate (EDTA), or a combination of both. Lysozyme is an enzyme that is present in egg white and in secretions such as tears and saliva, and which digests the polymeric compounds that give the cell wall its rigidity. EDTA, on the other hand, removes magnesium ions that are essential for preserving the overall structure of the cell envelope, and also inhibits cellular enzymes that could degrade DNA. Under some conditions, weakening the cell wall with lysozyme or EDTA is sufficient to cause bacterial cells to burst, but usually a detergent such as sodium dodecyl sulphate (SDS) is also added. Detergents aid the process of lysis by removing lipid molecules and thereby cause disruption of the cell membranes.

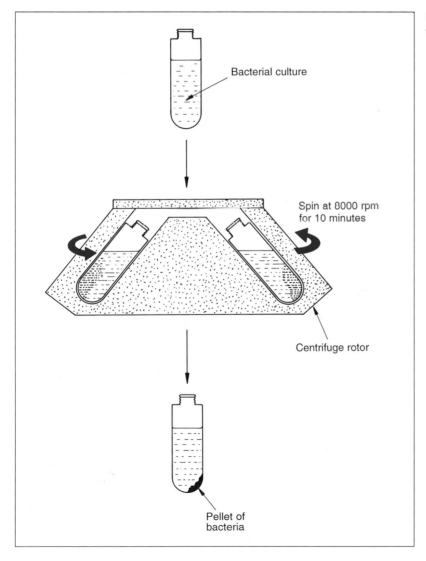

Figure 3.3 Harvesting bacteria by centrifugation.

Bacterial culture

Spin at 8000 rpm for 10 minutes

Centrifuge rotor

Pellet of bacteria

Figure 3.4 Preparation of a cell extract. (a) Cell lysis. (b) Centrifugation of the cell extract to remove insoluble debris.

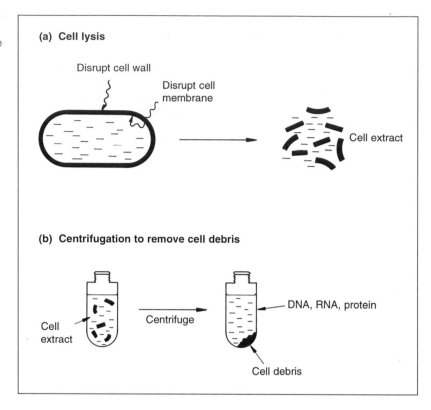

Having lysed the cells, the final step in preparation of a cell extract is removal of insoluble cell debris. Components such as partially digested cell wall fractions can be pelleted by centrifugation (Figure 3.4(b)), leaving the cell extract as a reasonably clear supernatant.

3.1.3 Purification of DNA from a cell extract

In addition to DNA, a bacterial cell extract will contain significant quantities of protein and RNA. A variety of procedures can be used to remove these contaminants, leaving the DNA in a pure form.

The standard way to deproteinize a cell extract is to add phenol or a 1:1 mixture of phenol and chloroform. These organic solvents precipitate proteins but leave the nucleic acids (DNA and RNA) in aqueous solution. The result is that if the cell extract is mixed gently with the solvent, and the layers then separated by centrifugation, precipitated protein molecules are left as a white coagulated mass at the interface between the aqueous and organic layers (Figure 3.5). The aqueous solution of nucleic acids can then be removed with a pipette.

With some cell extracts the protein content is so great that a sin-

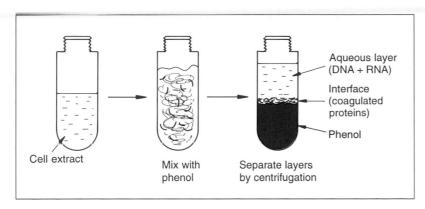

Figure 3.5 Removal of protein contaminants by phenol extraction.

Labels in figure: Cell extract; Mix with phenol; Separate layers by centrifugation; Aqueous layer (DNA + RNA); Interface (coagulated proteins); Phenol

gle phenol extraction is not sufficient to purify completely the nucleic acids. This problem could be solved by carrying out several phenol extractions one after the other, but this is undesirable as each mixing and centrifugation step results in a certain amount of breakage of the DNA molecules. The answer is to treat the cell extract with a **protease** such as Pronase or Proteinase K before phenol extraction. These enzymes break polypeptides down into smaller units, which are more easily removed by phenol.

Some RNA molecules, especially messenger RNA (mRNA), are removed by phenol treatment, but most remain with the DNA in the aqueous layer. The only effective way to remove the RNA is with the enzyme **ribonuclease**, which rapidly degrades these molecules into ribonucleotide subunits.

3.1.4 Concentration of DNA samples

Often a successful preparation results in a very thick solution of DNA that does not need to be concentrated any further. However, dilute solutions will sometimes be obtained and it is important to consider methods for increasing the DNA concentration.

The most frequently used method of concentration is **ethanol precipitation**. In the presence of salt (strictly speaking, monovalent cations such as Na^+), and at a temperature of $-20°C$ or less, absolute ethanol efficiently precipitates polymeric nucleic acids. With a thick solution of DNA the ethanol can be layered on top of the sample, causing molecules to precipitate at the interface. A spectacular trick is to push a glass rod through the ethanol into the DNA solution. When the rod is removed, DNA molecules will adhere and be pulled out of the solution in the form of a long fibre (Figure 3.6(a)). Alternatively, if ethanol is mixed with a dilute DNA solution, the precipitate can be collected by centrifugation (Figure 3.6(b)), and then redissolved in an appropriate volume of water. Ethanol precipitation has the added advantage of leaving in solution short-chain and monomeric nucleic acid components.

Figure 3.6 Collecting DNA by ethanol precipitation. (a) Absolute ethanol is layered on top of a concentrated solution of DNA. Fibres of DNA can be withdrawn with a glass rod. (b) For less concentrated solutions ethanol is added (at a ratio of 2.5 volumes of absolute ethanol to 1 volume of DNA solution) and precipitated DNA collected by centrifugation.

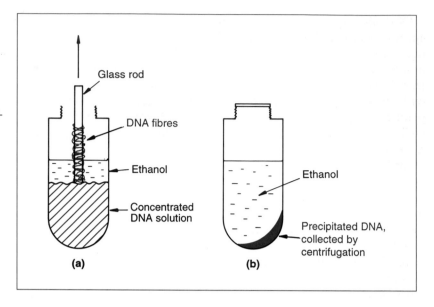

Ribonucleotides produced by ribonuclease treatment are therefore lost at this stage.

3.1.5 Measurement of DNA concentration

It is crucial to know exactly how much DNA is present in a solution when carrying out a gene cloning experiment. Fortunately, DNA concentrations can be accurately measured by **ultraviolet absorbance spectrophotometry**. The amount of ultraviolet radiation absorbed by a solution of DNA is directly proportional to the amount of DNA in the sample. Usually absorbance is measured at 260 nm, at which wavelength an absorbance (A_{260}) of 1.0 corresponds to 50 µg of double-stranded DNA per ml.

Ultraviolet absorbance can also be used to check the purity of a DNA preparation. With a pure sample of DNA the ratio of the absorbances at 260 nm and 280 nm (A_{260}/A_{280}) is 1.8. Ratios of less than 1.8 indicate that the preparation is contaminated, either with protein or with phenol.

3.1.6 Preparation of total cell DNA from organisms other than bacteria

Bacteria are not the only organisms from which DNA may be required. Total cell DNA from, for example, plants or animals will be needed if the aim of the genetic engineering project is to clone genes from these organisms. Although the basic steps in DNA purification are the same whatever the organism, some modifications may have to be introduced to take account of the special features of the cells being used.

Obviously growth of cells in liquid medium may not always be appropriate, even though plant and animal cell cultures are becoming increasingly important in biology. The major modifications, however, are likely to be needed at the cell breakage stage. The chemicals used for disrupting bacterial cells will not usually work with other organisms: lysozyme, for example, has no effect on plant cells. Specific degradative enzymes are available for most cell wall types, but often physical techniques, such as grinding frozen material with a mortar and pestle, will be more efficient. On the other hand, most animal cells have no cell wall at all, and can be lysed simply by treating with detergent.

Another important consideration is the biochemical content of the cells from which DNA is being extracted. With most bacteria the main biochemicals present in a cell extract are protein, DNA and RNA, so phenol extraction and/or protease treatment, followed by removal of RNA with ribonuclease, leaves a pure DNA sample. These treatments may not, however, be sufficient to give pure DNA if the cells also contain significant quantities of other biochemicals. Plant tissues are particularly difficult in this respect as they often contain large amounts of carbohydrates that are not removed by phenol extraction. Instead a different approach must be used. One method makes use of a detergent called cetyltrimethylammonium bromide (CTAB), which forms an insoluble complex with nucleic acids. When CTAB is added to a plant cell extract the nucleic acid–CTAB complex precipitates, leaving carbohydrate, protein and other contaminants in the supernatant (Figure 3.7). The precipitate is then collected by centrifugation and resuspended in 1 M NaCl, which causes the complex to break down. The nucleic acids can now be concentrated by ethanol precipitation and the RNA removed by ribonuclease treatment.

A second method makes use of the fact that nucleic acid molecules, unlike most of the contaminants in a cell extract, have relatively strong negative charges. This means that nucleic acids bind to positively charged surfaces, for instance to the particles in an anion-exchange chromatography resin (Figure 3.8(a)). One possibility would be to add the resin directly to the cell extract, just as in the CTAB procedure, but it is more convenient to use a chromatography column. The resin is placed in the column and the cell extract added (Figure 3.8(b)). Nucleic acids bind to the resin and are retained in the column, whereas the neutral and positively charged contaminants pass straight through. After washing away the last contaminants, the nucleic acids are recovered by adding a high-salt solution, which destabilizes the electrostatic interactions between the nucleic acid molecules and the resin.

Art Center College of Design
Library
1700 Lida Street
Pasadena, Calif. 91103

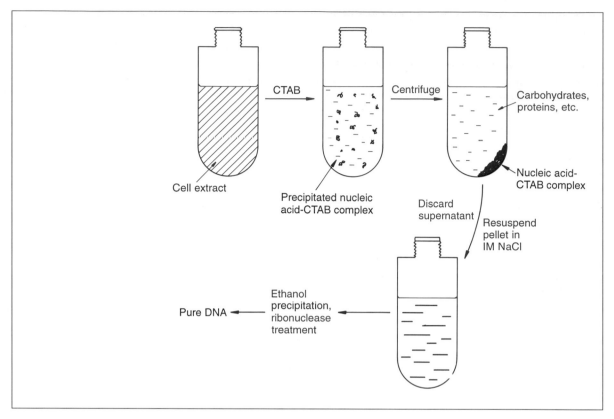

Figure 3.7 The CTAB method for purification of plant DNA.

3.2 PREPARATION OF PLASMID DNA

Purification of plasmids from a culture of bacteria involves the same general strategy as preparation of total cell DNA. A culture of cells, containing plasmids, is grown in liquid medium, harvested, and a cell extract prepared. The extract is deproteinized, the RNA removed, and the DNA probably concentrated by ethanol precipitation. However, there is an important distinction between plasmid purification and preparation of total cell DNA: in a plasmid preparation it is always necessary to separate the plasmid DNA from the large amount of bacterial chromosomal DNA that is also present in the cells.

Separating the two types of DNA can be very difficult, but is nonetheless essential if the plasmids are to be used as cloning vehicles. The presence of the smallest amount of contaminating bacterial DNA in a gene cloning experiment may easily lead to undesirable results. Fortunately, several methods are available for removal of bacterial DNA during plasmid purification, and the use

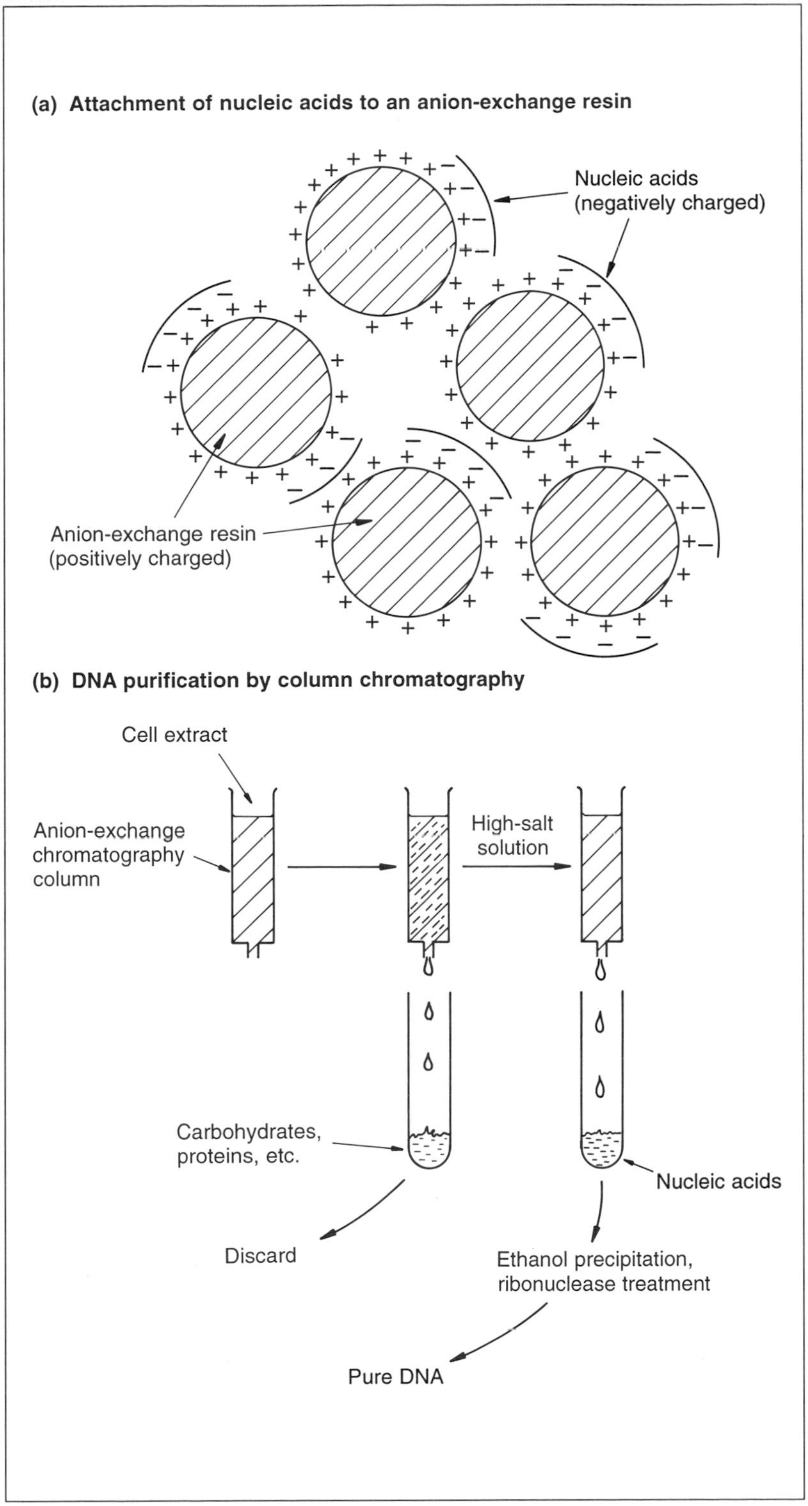

(a) Attachment of nucleic acids to an anion-exchange resin

Nucleic acids (negatively charged)

Anion-exchange resin (positively charged)

(b) DNA purification by column chromatography

Cell extract

Anion-exchange chromatography column

High-salt solution

Carbohydrates, proteins, etc.

Discard

Nucleic acids

Ethanol precipitation, ribonuclease treatment

Pure DNA

Figure 3.8 The use of an anion-exchange chromatography resin in DNA purification. (a) Negatively charged nucleic acid molecules bind to the positive charges on the surface of the particles in an anion-exchange resin. (b) DNA is purified by column chromatography.

of these methods, individually or in combination, can result in isolation of very pure plasmid DNA.

These methods are based on the several physical differences between plasmid DNA and bacterial DNA, the most obvious of which is size. The largest plasmids are only 8% of the size of the *E. coli* chromosome, and most are much smaller than this. Techniques that can separate small DNA molecules from large ones should therefore effectively purify plasmid DNA.

In addition to size, plasmids and bacterial DNA differ in **conformation**. When applied to a polymer such as DNA, the term conformation refers to the overall spatial configuration of the molecule, with the two simplest conformations being linear and circular. Plasmids and the bacterial chromosome are circular, but during preparation of the cell extract the chromosome will always be broken to give linear fragments. A method for separating circular from linear molecules will therefore result in pure plasmids.

3.2.1 Separation on the basis of size

The usual stage at which to attempt size fractionation is during preparation of the cell extract. If the cells are lysed under very carefully controlled conditions, then only a minimal amount of chromosomal DNA breakage will occur. The resulting DNA fragments are still very large, much larger than the plasmids, and can be removed with the cell debris by centrifugation. This process is aided by the fact that the bacterial chromosome is physically attached to the cell envelope, and fragments will almost certainly sediment with the cell debris if these attachments are not broken.

Cell disruption must therefore be performed very gently to prevent wholesale breakage of the bacterial DNA. For *E. coli* and related species, controlled lysis is performed as shown in Figure 3.9. Treatment with EDTA and lysozyme is carried out in the presence of sucrose, which prevents the cells from bursting straight away. Instead, **sphaeroplasts** are formed, partially wall-less cells that retain an intact cytoplasmic membrane. Cell lysis is now induced by adding a non-ionic detergent such as Triton X-100 (ionic detergents, such as SDS, cause chromosomal breakage). This method causes very little breakage of the bacterial DNA, so centrifugation will now leave a **cleared lysate**, consisting almost entirely of plasmid DNA.

A cleared lysate will, however, invariably retain some chromosomal DNA. Furthermore, if the plasmids themselves are large molecules, then they may also sediment with the cell debris. Size fractionation is therefore rarely sufficient on its own, and we must consider alternative ways of removing the bacterial DNA contaminants

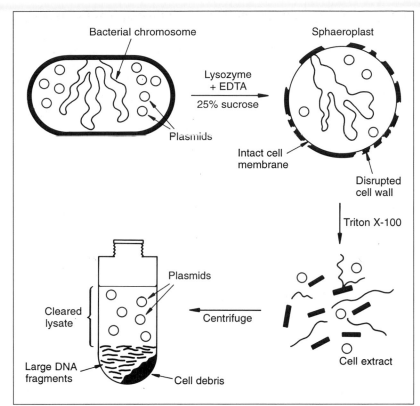

Figure 3.9 Preparation of a cleared lysate.

3.2.2 Separation on the basis of conformation

Before considering the ways in which conformational differences between plasmids and bacterial DNA can be used to separate the two types of DNA, we must look more closely at the overall structure of plasmid DNA. It is not strictly correct to say that plasmids have a circular conformation, because double-stranded DNA circles can in fact take up one of two quite distinct configurations. Most plasmids exist in the cell as **supercoiled** molecules (Figure 3.10(a)). Supercoiling occurs because the double helix of the plasmid DNA is partially unwound during the plasmid replication process by enzymes called topoisomerases (p. 59). The supercoiled conformation can be maintained only if both polynucleotide strands are intact, hence the more technical name of **covalently closed-circular (ccc) DNA**. If one of the polynucleotide strands is broken, then the double helix will revert to its normal, **relaxed** state, and the plasmid will take on the alternative conformation, called **open-circular (oc)** (Figure 3.10(b)).

Supercoiling is important in plasmid preparation because supercoiled molecules can be fairly easily separated from non-supercoiled DNA. Two different methods are commonly used.

Figure 3.10 Two conformations of circular double-stranded DNA. (a) Supercoiled: both strands are intact. (b) Open-circular: one or both strands are nicked.

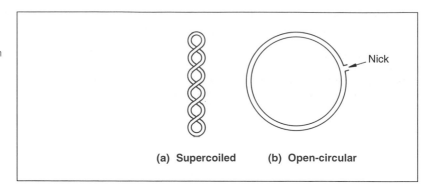

Both can purify plasmid DNA from crude cell extracts, though in practice best results are obtained if a cleared lysate is first prepared.

(a) Alkaline denaturation The basis to this technique is that there is a narrow pH range at which non-supercoiled DNA is **denatured**, whereas supercoiled plasmids are not. If sodium hydroxide is added to a cell extract or cleared lysate, so that the pH is adjusted to 12.0–12.5, then the hydrogen bonding in non-supercoiled DNA molecules is broken, causing the double helix to unwind and the two polynucleotide chains to separate (Figure 3.11). If acid is now added, these denatured bacterial DNA strands

Figure 3.11 Plasmid purification by the alkaline denaturation method.

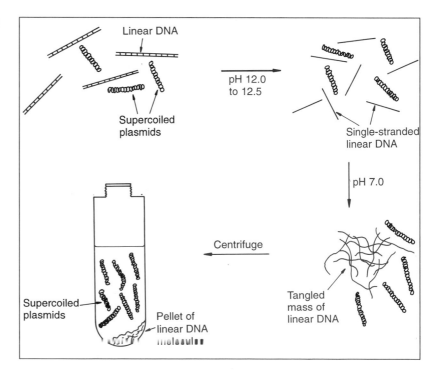

will reaggregate into a tangled mass. The insoluble network can be pelleted by centrifugation, leaving pure plasmid DNA in the supernatant. An additional advantage of this procedure is that, under some circumstances (specifically, cell lysis by SDS and neutralization with sodium acetate), most of the protein and RNA also becomes insoluble and can be removed by the centrifugation step. Phenol extraction and ribonuclease treatment may therefore not be needed if the alkaline denaturation method is used.

(b) Ethidium bromide – caesium chloride density gradient centrifugation   This is a specialized version of the more general technique of equilibrium or **density gradient centrifugation**. A density gradient is produced by centrifuging a solution of caesium chloride (CsCl) at a very high speed (Figure 3.12(a)). The gradient develops because a high centrifugal force pulls the caesium and chloride ions towards the bottom of the tube. Their downward migration is counterbalanced by diffusion, so a concentration gradient is set up, with the CsCl density greater towards the bottom of the tube.

Macromolecules present in the CsCl solution when it is centrifuged will form bands at distinct points in the gradient (Figure 3.12(b)). Exactly where a particular molecule bands depends on its **buoyant density**. DNA has a buoyant density of about $1.7 \, \text{g/cm}^3$, and therefore migrates to the point in the gradient where the CsCl density is also $1.7 \, \text{g/cm}^3$. In contrast, protein molecules have much lower buoyant densities, and so float at the top of the tube, whereas RNA forms a pellet at the bottom (Figure 3.12(b)). Density

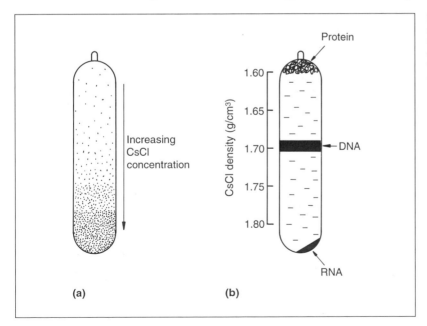

Figure 3.12 CsCl density gradient centrifugation. (a) A CsCl density gradient produced by high-speed centrifugation. (b) Separation of protein, DNA and RNA in a density gradient.

gradient centrifugation can therefore separate DNA, RNA and protein and is an alternative to phenol extraction and ribonuclease treatment for DNA purification.

More importantly, density gradient centrifugation in the presence of **ethidium bromide (EtBr)** can be used to separate supercoiled DNA from non-supercoiled molecules. EtBr binds to DNA molecules by intercalating between adjacent base pairs, causing partial unwinding of the double helix (Figure 3.13). This unwinding results in a decrease in the buoyant density, by as much as 0.125 g/cm^3 for linear DNA. However, supercoiled DNA, with no free ends, has very little freedom to unwind, and can only bind a limited amount of EtBr. The decrease in buoyant density of a supercoiled molecule is therefore much less, only about 0.085 g/cm^3. As a consequence, supercoiled molecules will band in an EtBr–CsCl gradient at a different position to linear and open-circular DNA (Figure 3.14(a)).

EtBr–CsCl density gradient centrifugation is a very efficient method for obtaining pure plasmid DNA. When a cleared lysate is subjected to this procedure, plasmids band at a distinct point, separated from the linear bacterial DNA, with the protein floating on

Figure 3.13 Partial unwinding of the DNA double helix by EtBr intercalation between adjacent base pairs. The normal DNA molecule shown on the left is partially unwound by taking up four EtBr molecules, resulting in the 'stretched' structure on the right.

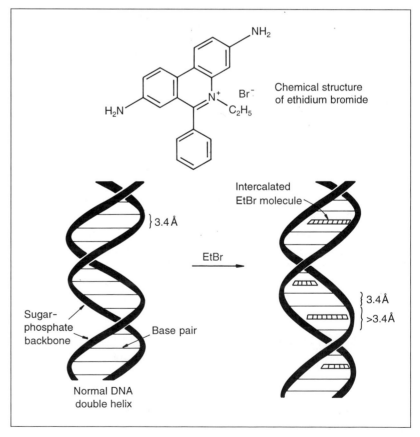

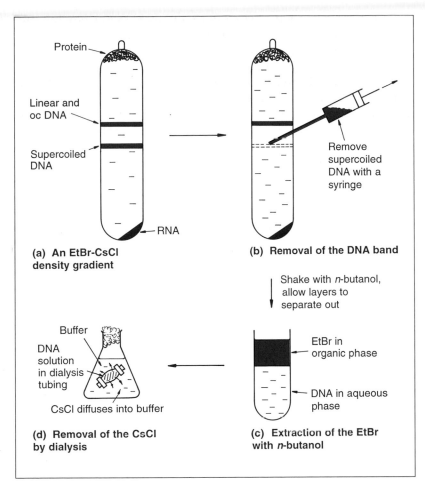

Figure 3.14 Purification of plasmid DNA by EtBr–CsCl density gradient centrifugation.

the top of the gradient and RNA pelleted at the bottom. The position of the DNA bands can be seen by shining ultraviolet radiation on the tube, which causes the bound EtBr to fluoresce. The pure plasmid DNA is removed by puncturing the side of the tube and withdrawing a sample with a syringe (Figure 3.14(b)). The EtBr bound to the plasmid DNA is extracted with *n*-butanol (Figure 3.14(c)) and the CsCl removed by dialysis (Figure 3.14(d)). The resulting plasmid preparation is virtually 100% pure and ready for use as a cloning vehicle.

3.2.3 Plasmid amplification

Preparation of plasmid DNA can be hindered by the fact that plasmids make up only a small proportion of the total DNA in the bacterial cell. The yield of DNA from a bacterial culture may therefore be disappointingly low. **Plasmid amplification** offers a means of increasing this yield.

The aim of amplification is to increase the copy number of a plasmid. Some **multicopy plasmids** (those with copy numbers of 20 or more) have the useful property of being able to replicate in the absence of protein synthesis. This contrasts with the main bacterial chromosome, which cannot replicate under these conditions. This property can be made use of during the growth of a bacterial culture for plasmid DNA purification. After a satisfactory cell density has been reached, an inhibitor of protein synthesis (for example, chloramphenicol) is added, and the culture incubated for a further 12 hours. During this time the plasmid molecules continue to replicate, even though chromosomal replication and cell division are blocked (Figure 3.15). The result is that plasmid copy numbers of several thousand may be attained. Amplification is therefore a very efficient way of increasing the yield of multicopy plasmids.

Figure 3.15 Plasmid amplification.

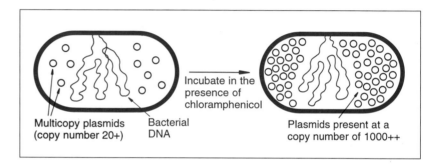

Multicopy plasmids (copy number 20+)   Bacterial DNA

Incubate in the presence of chloramphenicol

Plasmids present at a copy number of 1000++

3.3 PREPARATION OF BACTERIOPHAGE DNA

The key difference between phage DNA purification and the preparation of either total cell DNA or plasmid DNA, is that for phages the starting material is not normally a cell extract. This is because bacteriophage particles can be obtained in large numbers from the extracellular medium of an infected bacterial culture. When such a culture is centrifuged, the bacteria are pelleted, leaving the phage particles in suspension (Figure 3.16). The phage particles are then collected from the suspension and their DNA extracted by a single deproteinization step to remove the phage capsid.

This overall process is rather more straightforward than the procedure used to prepare total cell or plasmid DNA. Nevertheless, successful purification of significant quantities of phage DNA is subject to several pitfalls. The main difficulty, especially with λ, is growing an infected culture in such a way that the extracellular phage titre (meaning the number of phage particles per ml of culture) is sufficiently high. In practical terms, the maximum titre that

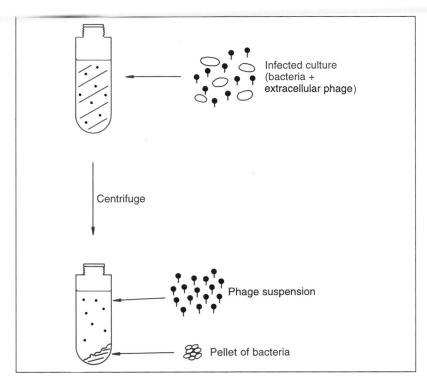

Figure 3.16 Preparation of a phage suspension from an infected culture of bacteria.

can reasonably be expected for λ is 10^{10} per ml; yet 10^{10} λ particles will yield only 500 ng of DNA. Large culture volumes, in the range of 500 to 1000 ml are therefore needed if substantial quantities of λ DNA are to be obtained.

3.3.1 Growth of cultures to obtain a high λ titre

Growing a large-volume culture is no problem (bacterial cultures of 50 litres and over are common in biotechnology), but obtaining the maximum phage titre requires a certain amount of skill. The naturally occurring λ phage is lysogenic (p. 20), and an infected culture consists mainly of cells carrying the prophage integrated into the bacterial DNA (Figure 2.7). The extracellular λ titre is extremely low under these circumstances.

To get a high yield of extracellular λ, the culture must be **induced**, so that all the bacteria enter the lytic phase of the infection cycle, resulting in cell death and release of λ particles into the medium. Induction is normally very difficult to control, but most laboratory strains of λ carry a **temperature-sensitive (ts) mutation** in the *c*I gene. This is one of the genes that are responsible for maintaining the phage in the integrated state. If inactivated by a mutation, the *c*I gene will no longer function correctly and the switch to lysis will occur. In the *c*I*ts* mutation, the *c*I gene is functional at 30°C, at which temperature normal lysogeny can occur. But at 42°C, the *cIts* gene product does not work properly, and lysogeny cannot be maintained. A culture of *E. coli* infected with λ *cIts* can therefore be induced to produce extracellular phage by transferring from 30°C to 42°C (Figure 3.17).

Figure 3.17 Induction of a λ.*cIts* lysogen by transferring from 30°C to 42°C.

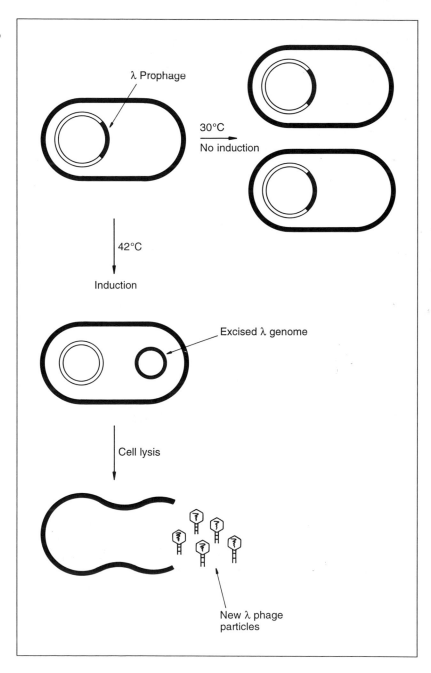

3.3.2 Preparation of non-lysogenic λ phage

Although most λ strains are lysogenic, many cloning vectors derived from λ are modified, by deletions of the *cI* and other genes, so that lysogeny will never occur. These phages cannot integrate into the bacterial genome and can infect cells only by a lytic cycle (p. 19).

   With these phages the key to obtaining a high titre lies in the way in which the culture is grown, in particular the stage at which the cells are infected by adding phage particles. If phages are added before the cells are dividing at their maximal rate, then all the cells will be lysed very quickly, resulting in a low titre (Figure 3.18(a)). On the other hand, if the cell density is too high when the phages are added, then the culture will never be completely lysed, and again the phage titre will be low (Figure 3.18(b)). The ideal situation is when the age of the culture, and the size of the phage

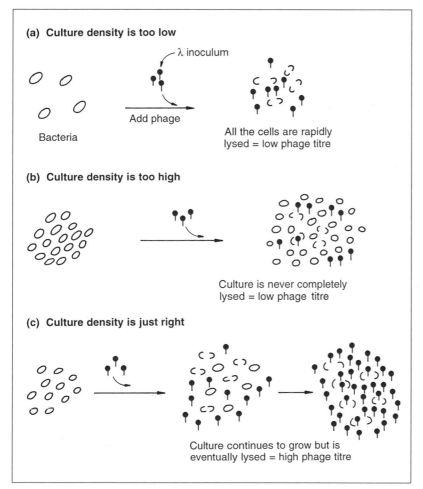

(a) **Culture density is too low**

λ inoculum

Bacteria

Add phage

All the cells are rapidly
lysed = low phage titre

(b) **Culture density is too high**

Culture is never completely
lysed = low phage titre

(c) **Culture density is just right**

Culture continues to grow but is
eventually lysed = high phage titre

Figure 3.18 Achieving the right balance between culture age and inoculum size when preparing a sample of a non-lysogenic phage.

inoculum, are balanced such that the culture continues to grow, but eventually all the cells are infected and lysed (Figure 3.18(c)). As can be imagined, skill and experience are needed to judge the matter to perfection.

3.3.3 Collection of phage from an infected culture

The remains of lysed bacterial cells, along with any intact cells that are inadvertently left over, can be removed from an infected culture by centrifugation, leaving the phage particles in suspension (Figure 3.16). The problem now is to reduce the size of the suspension to 5 ml or less, a manageable size for DNA extraction.

Phage particles are so small that they are pelleted only by very high speed centrifugation. Collection of phages is therefore usually achieved by precipitation with **polyethylene glycol (PEG)**. This is a long-chain polymeric compound which, in the presence of salt, absorbs water, thereby causing macromolecular assemblies such as phage particles to precipitate. The precipitate can then be collected by centrifugation, and redissolved in a suitably small volume (Figure 3.19).

Figure 3.19 Collection of phage particles by polyethylene glycol (PEG) precipitation.

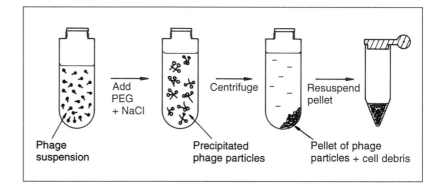

Phage suspension — Add PEG + NaCl → Precipitated phage particles — Centrifuge → — Resuspend pellet → Pellet of phage particles + cell debris

3.3.4 Purification of DNA from λ phage particles

Deproteinization of the redissolved PEG precipitate is sometimes sufficient to extract pure phage DNA, but usually λ phages are subjected to an intermediate purification step. This is necessary because the PEG precipitate will also contain a certain amount of bacterial debris, possibly including unwanted cellular DNA. These contaminants can be separated from the λ particles by CsCl density gradient centrifugation. λ particles band in a CsCl gradient at 1.45 to 1.50 g/cm³ (Figure 3.20), and can be withdrawn from the gradient just as described previously for DNA bands (p. 43 and Figure 3.14). Removal of CsCl by dialysis leaves a pure phage preparation

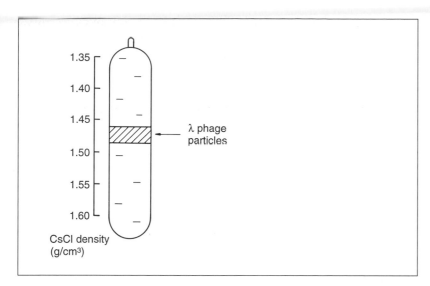

1.35

1.40

1.45

λ phage
particles

1.50

1.55

1.60

CsCl density
(g/cm³)

Figure 3.20 Purification of λ phage
particles by CsCl density gradient
centrifugation.

from which the DNA can be extracted by either phenol or protease
treatment to digest the phage protein coat.

3.3.5 Purification of M13 DNA causes few problems

Most of the differences between the M13 and λ infection cycles are
to the advantage of the molecular biologist wishing to prepare M13
DNA. First, the double-stranded replicative form of M13 (p. 25),
which behaves like a high copy number plasmid, is very easily
purified by the standard procedures for plasmid preparation. A
cell extract is prepared from cells infected with M13, and the
replicative form separated from bacterial DNA by, for example,
EtBr–CsCl density gradient centrifugation.

However, the single-stranded form of the M13 genome, contained in the extracellular phage particles, is frequently required.
In this respect, the big advantage compared with λ is that high
titres of M13 are very easy to obtain. As infected cells continually
secrete M13 particles into the medium (Figure 2.8), with lysis never
occurring, a high M13 titre is achieved simply by growing the
infected culture to a high cell density. In fact titres of 10^{12} per ml
and above are quite easy to obtain without any special tricks being
used. Such high titres mean that significant amounts of single-
stranded M13 DNA can be prepared from cultures of small volume – 5 ml or less. Furthermore, as the infected cells are not lysed,
there is no problem with cell debris contaminating the phage suspension. Consequently the CsCl density gradient centrifugation
step, needed for λ phage preparation, is rarely required with M13.

In summary, single-stranded M13 DNA preparation involves
growth of a small volume of infected culture, centrifugation to

pellet the bacteria, precipitation of the phage particles with PEG, phenol extraction to remove the phage protein coats, and ethanol precipitation to concentrate the resulting DNA (Figure 3.21).

Figure 3.21 Preparation of M13 DNA from an infected culture of bacteria.

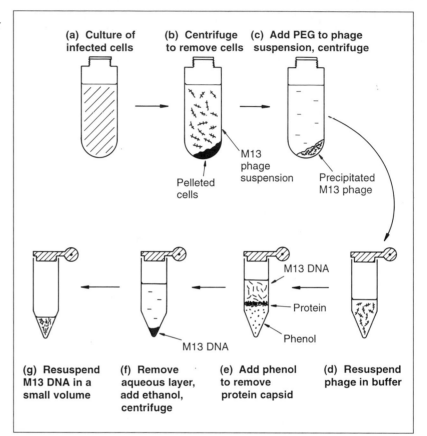

(a) **Culture of infected cells**

(b) **Centrifuge to remove cells**

(c) **Add PEG to phage suspension, centrifuge**

Pelleted cells

M13 phage suspension

Precipitated M13 phage

M13 DNA

Protein

Phenol

M13 DNA

(g) **Resuspend M13 DNA in a small volume**

(f) **Remove aqueous layer, add ethanol, centrifuge**

(e) **Add phenol to remove protein capsid**

(d) **Resuspend phage in buffer**

FURTHER READING

Marmur, J. (1961) A procedure for the isolation of deoxyribonucleic acid from micro-organisms. *Journal of Molecular Biology*, **3**, 208–18 – total cell DNA preparation.

Rogers, S. O. and Bendich, A. J. (1985) Extraction of DNA from milligram amounts of fresh, herbarium and mummified plant tissues. *Plant Molecular Biology*, **5**, 69–76 – the CTAB method.

Birnboim, H. C. and Doly, J. (1979) A rapid alkaline extraction procedure for screening recombinant plasmid DNA. *Nucleic Acids Research*, **7**, 1513–23 – a method for preparing plasmid DNA.

Radloff, R., Bauer, W. and Vinograd, J. (1967) A dye–buoyant-density method for the detection and isolation of closed circular duplex DNA. *Proceedings of the National Academy of Sciences, USA*, **57**, 1514–21 – the original description of ethidium bromide density gradient centrifugation.

Clewell, D. B. (1972) Nature of ColE1 plasmid replication in *Escherichia coli* in the presence of chloramphenicol. *Journal of Bacteriology*, **110**, 667–76 – the biological basis to plasmid amplification.

Yamamoto, K. R., Alberts, B. M., Benzinger, R., Lawhorne, L. and Trieber, G. (1970) Rapid bacteriophage sedimentation in the presence of polyethylene glycol and its application to large scale virus preparation. *Virology*, **40**, 734–44 – preparation of λ DNA.

Zinder, N. D. and Boeke, J. D. (1982) The filamentous phage (Ff) as vectors for recombinant DNA. *Gene*, **19**, 1–10 – methods for phage growth and DNA preparation.

4 Manipulation of purified DNA

Once pure samples of DNA have been prepared, the next step in a gene cloning experiment is construction of the recombinant DNA molecule (Figure 1.1). To produce this recombinant molecule, the vector, as well as the DNA to be cloned, must be cut at specific points and then joined together in a controlled manner. Cutting and joining are two examples of DNA manipulative techniques, a wide variety of which have been developed over the last few years. As well as being cut and joined, DNA molecules can be shortened, lengthened, copied into RNA or into new DNA molecules, and modified by the addition or removal of specific chemical groups. These manipulations, all of which can be carried out in the test-tube, provide the foundation not only for gene cloning, but also for basic studies into DNA biochemistry, gene structure, and the control of gene expression.

Almost all DNA manipulative techniques make use of purified enzymes. Within the cell these enzymes participate in essential processes such as DNA replication and transcription, breakdown of unwanted or foreign DNA (for example, invading virus DNA), repair of mutated DNA, and **recombination** between different DNA molecules. After purification from cell extracts, many of these enzymes can be persuaded to carry out their natural reactions, or something closely related to them, under artificial conditions. Although these enzymatic reactions are often quite straightforward, most are absolutely impossible to perform by standard chemical methods. Purified enzymes are therefore crucial to genetic engineering and an important industry has sprung up around their preparation, characterization and marketing. Commercial suppliers of high-purity enzymes provide an essential service to the molecular biologist.

The cutting and joining manipulations that underlie gene cloning are carried out by enzymes called restriction endonucleases (for cutting) and ligases (for joining). Most of this chapter will be concerned with the ways in which these two types of enzyme are used. Firstly, however, we must consider the whole range of DNA manipulative enzymes, to see exactly what types of reaction can be performed. Many of these enzymes will be mentioned in later chapters when procedures that make use of them will be described.

4.1 THE RANGE OF DNA MANIPULATIVE ENZYMES

These enzymes can be grouped into five broad classes depending on the type of reaction that they catalyse.

1. **Nucleases** are enzymes that cut, shorten or degrade nucleic acid molecules.
2. **Ligases** join nucleic acid molecules together.
3. **Polymerases** make copies of molecules.
4. **Modifying enzymes** remove or add chemical groups.
5. **Topoisomerases** introduce or remove supercoils from covalently closed-circular DNA.

Before considering in detail each of these classes of enzyme, two points should be made. The first is that, although most enzymes can be assigned to a particular class, a few display multiple activities that span two or more classes. Most importantly, many polymerases combine their ability to make new DNA molecules with an associated DNA degradative (= nuclease) activity.

Secondly it should be appreciated that, as well as the DNA manipulative enzymes, many similar enzymes able to act on RNA are known. The ribonuclease used to remove contaminating RNA from DNA preparations (p. 33) is an example of such an enzyme. Although some RNA manipulative enzymes have applications in gene cloning and will be mentioned in later chapters, we will in general restrict our thoughts to those enzymes that act on DNA.

4.1.1 Nucleases

Nucleases degrade DNA molecules by breaking the phosphodiester bonds that link one nucleotide to the next in a DNA strand. There are two distinct kinds of nuclease (Figure 4.1).

1. **Exonucleases**, which remove nucleotides one at a time from the end of a DNA molecule.
2. **Endonucleases**, which are able to break internal phosphodiester bonds within a DNA molecule.

The main distinction between different exonucleases lies in the number of strands that are degraded when a double-stranded molecule is attacked. The enzyme called Bal31 (purified from the

Figure 4.1 The reactions catalysed by the two different kinds of nuclease. (a) An exonuclease, which removes nucleotides from the end of a DNA molecule. (b) An endonuclease, which breaks internal phosphodiester bonds.

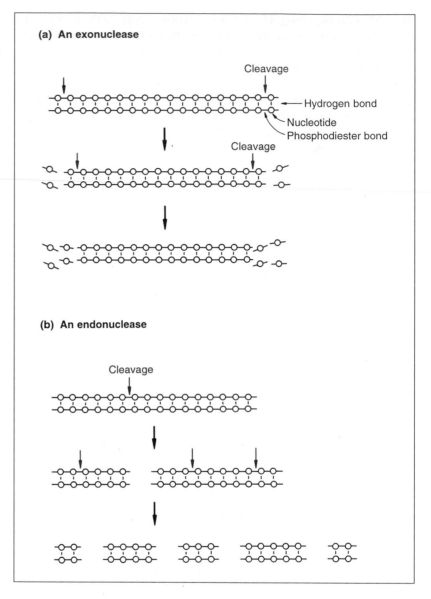

bacterium *Alteromonas espejiana*) is an example of an exonuclease that removes nucleotides from both strands of a double-stranded molecule (Figure 4.2(a)). The greater the length of time that Bal31 is allowed to act on a group of DNA molecules, the shorter the resulting DNA fragments will be. In contrast, enzymes such as *E. coli* exonuclease III degrade just one strand of a double-stranded molecule, leaving single-stranded DNA as the product (Figure 4.2(b)).

The same criterion can be used to classify endonucleases. S1 endonuclease (from the fungus *Aspergillus oryzae*) will cleave only single strands (Figure 4.3(a)), whereas deoxyribonuclease I (DNase

1), which is prepared from cow pancreas, cuts both single- and double-stranded molecules (Figure 4.3(b)). DNase I is non-specific in that it attacks DNA at any internal phosphodiester bond; the end result of prolonged DNase I action is therefore a mixture of mononucleotides and very short oligonucleotides. On the other hand, the special group of enzymes called restriction endonucleases cleave double-stranded DNA only at a limited number of specific recognition sites (Figure 4.3(c)). These important enzymes are described in detail on p. 60.

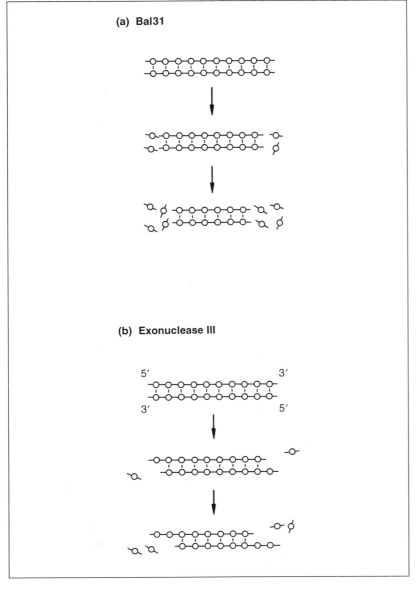

(a) **Bal31**

(b) **Exonuclease III**

Figure 4.2 The reactions catalysed by different types of exonuclease. (a) Bal31, which removes nucleotides from both strands of a double-stranded molecule. (b) Exonuclease III, which removes nucleotides only from the 3′-terminus (see p. 81 for a description of the differences between the 3′- and 5′-termini of a polynucleotide).

Figure 4.3 The reactions catalysed by different types of endonuclease. (a) S1 nuclease, which cleaves only single-stranded DNA, including single-stranded nicks in mainly double-stranded molecules. (b) DNase I, which cleaves both single- and double-stranded DNA. (c) A restriction endonuclease, which cleaves double-stranded DNA, but only at a limited number of sites.

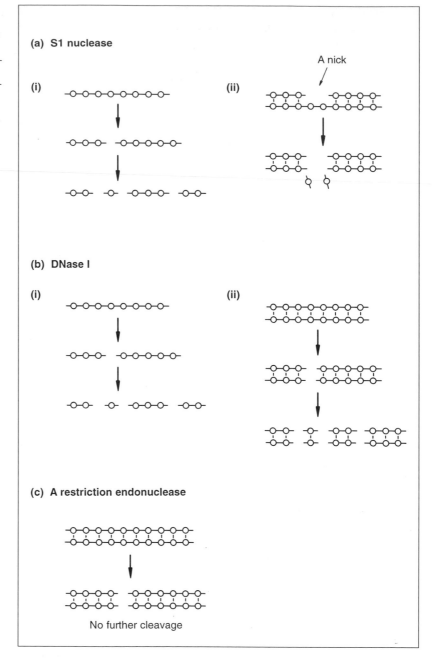

4.1.2 Ligases

In the cell the function of DNA ligase is to repair single-stranded breaks ('discontinuities') that arise in double-stranded DNA molecules during, for example, DNA replication. DNA ligases from most organisms will also join together two individual fragments of

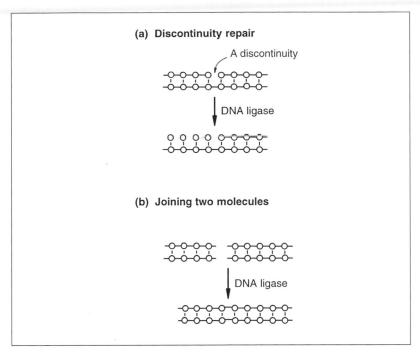

(a) Discontinuity repair

(b) Joining two molecules

Figure 4.4 The two reactions catalysed by DNA ligase. (a) Repair of a discontinuity – a missing phosphodiester bond in one stand of a double-stranded molecule. (b) Joining two molecules together.

double-stranded DNA (Figure 4.4). The role of these enzymes in construction of recombinant DNA molecules is described on p. 77.

4.1.3 Polymerases

DNA polymerases are enzymes that synthesize a new strand of DNA complementary to an existing DNA or RNA **template** (Figure 4.5(a)). Most polymerases can function only if the template possesses a double-stranded region which acts as a **primer** for initiation of polymerization.

Three types of DNA polymerase are used routinely in genetic engineering. The first is DNA polymerase I, which is prepared usually from *E. coli*. This enzyme attaches to a short single-stranded region (or '**nick**') in a mainly double-stranded DNA molecule, and then synthesizes a completely new strand, degrading the existing strand as it proceeds (Figure 4.5(b)). DNA polymerase I is therefore an example of an enzyme with a dual activity – DNA polymerization and DNA degradation.

In fact the polymerase and nuclease activities of DNA polymerase I are controlled by different parts of the enzyme molecule. The nuclease activity is contained in the first 323 amino acids of the polypeptide, so removal of this segment leaves a modified enzyme that retains the polymerase function but is unable to degrade DNA. This modified enzyme, called the **Klenow fragment**, can still synthesize a complementary DNA strand on a single-stranded template,

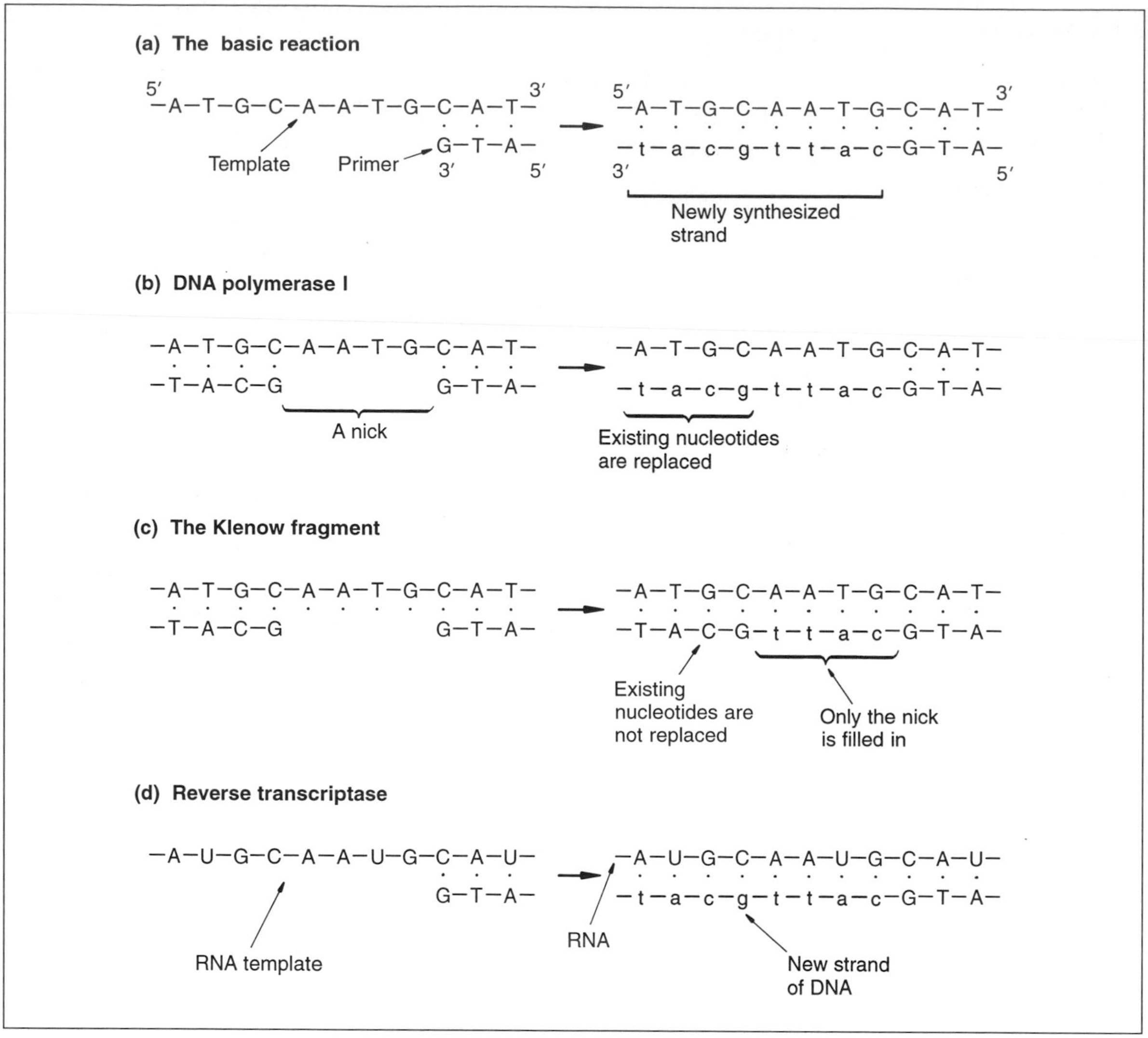

(a) The basic reaction

5'   3'   5'   3'
−A−T−G−C−A−A−T−G−C−A−T−   −A−T−G−C−A−A−T−G−C−A−T−
                                ⋅ ⋅ ⋅   ⋅ ⋅ ⋅
                             G−T−A−   −t−a−c−g−t−t−a−c−G−T−A−
Template   Primer   3'   5'   3'   5'

Newly synthesized
strand

(b) DNA polymerase I

−A−T−G−C−A−A−T−G−C−A−T−   −A−T−G−C−A−A−T−G−C−A−T−
⋅ ⋅ ⋅   ⋅ ⋅ ⋅   ⋅ ⋅ ⋅ ⋅ ⋅ ⋅ ⋅ ⋅ ⋅ ⋅ ⋅
−T−A−C−G   G−T−A−   −t−a−c−g−t−t−a−c−G−T−A−

A nick   Existing nucleotides
                                             are replaced

(c) The Klenow fragment

−A−T−G−C−A−A−T−G−C−A−T−   −A−T−G−C−A−A−T−G−C−A−T−
⋅ ⋅ ⋅   ⋅ ⋅ ⋅   ⋅ ⋅ ⋅ ⋅ ⋅ ⋅ ⋅ ⋅ ⋅ ⋅ ⋅
−T−A−C−G   G−T−A−   −T−A−C−G−t−t−a−c−G−T−A−

Existing   Only the nick
nucleotides are   is filled in
not replaced

(d) Reverse transcriptase

−A−U−G−C−A−A−U−G−C−A−U−   −A−U−G−C−A−A−U−G−C−A−U−
                                ⋅ ⋅ ⋅   ⋅ ⋅ ⋅ ⋅ ⋅ ⋅ ⋅ ⋅ ⋅ ⋅ ⋅
                             G−T−A−   −t−a−c−g−t−t−a−c−G−T−A−
                                                      RNA
RNA template   New strand
                                                                 of DNA

Figure 4.5 The reactions catalysed by DNA polymerases. (a) The basic reaction: a new DNA strand is synthesized in the 5' to 3' direction. (b) DNA polymerase I, which initially fills in nicks but then continues to synthesize a new strand, degrading the existing one as it proceeds. (c) The Klenow fragment, which only fills in nicks. (d) Reverse transcriptase, which uses a template of RNA.

but as it has no nuclease activity it cannot continue the synthesis once the nick is filled in (Figure 4.5(c)). Several other enzymes – natural polymerases and modified versions – are able to perform the same function. The major application of the Klenow fragment and these related polymerases is in DNA sequencing (p. 192).

The third type of DNA polymerase that is important in genetic engineering is **reverse transcriptase**, an enzyme involved in the replication of several kinds of virus. Reverse transcriptase is unique in that it uses as a template not DNA but RNA (Figure 4.5(d)). The ability of this enzyme to synthesize a DNA strand complementary to an RNA template is central to the technique called cDNA cloning (p. 167).

4.1.4 DNA modifying enzymes

There are numerous enzymes that modify DNA molecules by addition or removal of specific chemical groups. The most important are as follows.

1. **Alkaline phosphatase** (from *E. coli* or calf intestinal tissue) which removes the phosphate group present at the **5′-terminus** of a DNA molecule (Figure 4.6(a)).
2. **Polynucleotide kinase** (from *E. coli* infected with T4 phage) which has the reverse effect to alkaline phosphatase, adding phosphate groups on to free 5′-termini (Figure 4.6(b)).
3. **Terminal deoxynucleotidyl transferase** (from calf thymus tissue) which adds one or more deoxynucleotides on to the **3′-terminus** of a DNA molecule (Figure 4.6(c)).

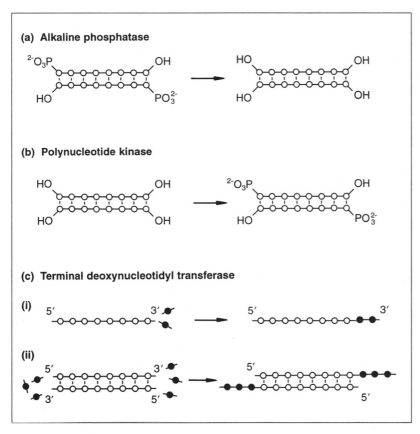

(a) Alkaline phosphatase

(b) Polynucleotide kinase

(c) Terminal deoxynucleotidyl transferase

(i)

(ii)

Figure 4.6 The reactions catalysed by DNA-modifying enzymes. (a) Alkaline phosphatase, which removes 5′-phosphate groups. (b) Polynucleotide kinase, which attaches 5′-phosphate groups. (c) Terminal deoxynucleotidyl transferase, which attaches nucleotides to the 3′-termini of polynucleotides in either (i) single-stranded or (ii) double-stranded molecules.

4.1.5 Topoisomerases

The final class of DNA manipulative enzymes are the topoisomerases, which are able to change the conformation of covalently closed-circular DNA (e.g. plasmid DNA molecules) by introducing or removing supercoils (p. 39). Although important in the study of DNA replication, topoisomerases have yet to find a real use in genetic engineering.

4.2 ENZYMES FOR CUTTING DNA – RESTRICTION ENDONUCLEASES

Gene cloning requires that DNA molecules be cut in a very precise and reproducible fashion. This is illustrated by the way in which the vector is cut during construction of a recombinant DNA molecule (Figure 4.7(a)). Each vector molecule must be cleaved at a single position, to open up the circle so that new DNA can be inserted; a molecule that is cut more than once will be broken into two or more separate fragments and will be of no use as a cloning vehicle. Furthermore, each vector molecule must be cut at exactly the same position on the circle – as will become apparent in later chapters, random cleavage is not satisfactory. It should be clear

Figure 4.7 The need for very precise cutting manipulations in a gene cloning experiment.

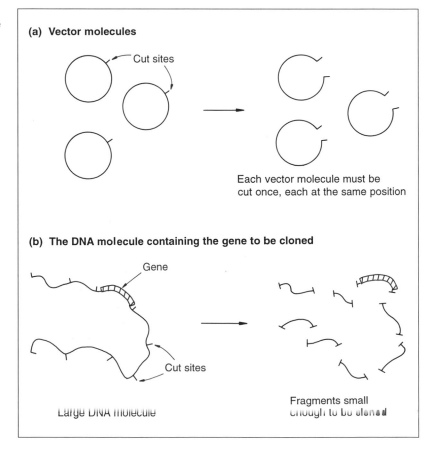

(a) Vector molecules

Cut sites

Each vector molecule must be cut once, each at the same position

(b) The DNA molecule containing the gene to be cloned

Gene

Cut sites

Large DNA molecule

Fragments small enough to be cloned

that a very special type of nuclease is needed to carry out this manipulation.

Often it will also be necessary to cleave the DNA that is to be cloned (Figure 4.7(b)). There are two reasons for this. Firstly, if the aim is to clone a single gene, which may consist of only 2 or 3 kb of DNA, then that gene will have to be cut out of the large (often greater than 80 kb) DNA molecules produced by skilful use of the preparative techniques described in Chapter 3. Secondly, large DNA molecules may have to be broken down simply to produce fragments small enough to be carried by the vector. Most cloning vectors exhibit a preference for DNA fragments that fall into a particular size range; M13-based vectors, for example, are very inefficient at cloning DNA molecules of more than 3 kb in length.

Purified restriction endonucleases allow the molecular biologist to cut DNA molecules in the precise, reproducible manner required for gene cloning. The discovery of these enzymes, which led to Nobel Prizes for W. Arber, H. Smith and D. Nathans in 1978, was one of the key breakthroughs in the development of genetic engineering.

4.2.1 The discovery and function of restriction endonucleases

The initial observation that led to the eventual discovery of restriction endonucleases was made in the early 1950s, when it was shown that some strains of bacteria are immune to bacteriophage infection, a phenomenon referred to as **host-controlled restriction**.

The mechanism of restriction is not very complicated, even though it took over 20 years to be fully understood. Restriction occurs because the bacterium produces an enzyme that degrades the phage DNA before it has time to replicate and direct synthesis of new phage particles (Figure 4.8(a)). The bacterium's own DNA, the destruction of which would of course be lethal, is protected from attack because it carries additional methyl groups that block the degradative enzyme action (Figure 4.8(b)). These degradative enzymes are called restriction endonucleases and are synthesized by many, perhaps all, species of bacteria; over 1200 different ones have now been characterized. Three different classes of restriction endonuclease are recognized, each distinguished by a slightly different mode of action. Types I and III are rather complex and have only a very limited role in genetic engineering. Type II restriction endonucleases, on the other hand, are the cutting enzymes that are so important in gene cloning.

Figure 4.8 The function of a restriction endonuclease in a bacterial cell. (a) Phage DNA is cleaved, but (b) bacterial DNA is not.

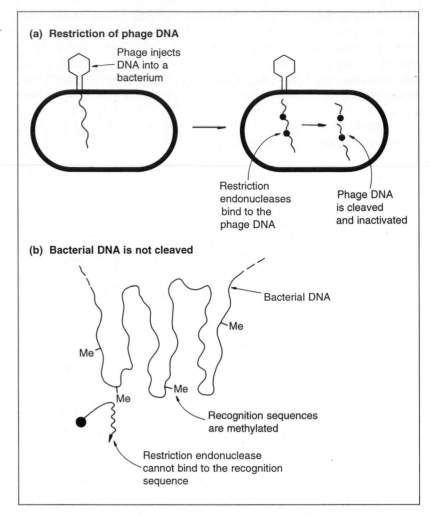

(a) **Restriction of phage DNA**

Phage injects DNA into a bacterium

Restriction endonucleases bind to the phage DNA

Phage DNA is cleaved and inactivated

(b) **Bacterial DNA is not cleaved**

Bacterial DNA

Me

Me

Me

Me

Recognition sequences are methylated

Restriction endonuclease cannot bind to the recognition sequence

4.2.2 Type II restriction endonucleases cut DNA at specific nucleotide sequences

The central feature of Type II restriction endonucleases (which will be referred to simply as 'restriction endonucleases' from now on) is that each enzyme has a specific recognition sequence at which it cuts a DNA molecule. A particular enzyme will cleave DNA at the recognition sequence and nowhere else. For example, the restriction endonuclease called *Pvu*I (isolated from *Proteus vulgaris*) cuts DNA only at the hexanucleotide CGATCG. In contrast, a second enzyme from the same bacterium, called *Pvu*II, cuts at a different hexanucleotide, in this case CAGCTG.

Many restriction endonucleases recognize hexanucleotide target sites, but others cut at four, five or even eight nucleotide sequences.

*Sau*3A (from *Staphylococcus aureus* strain 3A) recognizes GATC, and *Alu*I (*Arthrobacter luteus*) cuts at AGCT. There are also examples of restriction endonucleases with degenerate recognition sequences, meaning that they cut DNA at any one of a family of related sites. *Hin*fI (*Haemophilus influenzae* strain R$_f$), for instance, recognizes GANTC, so cuts at GAATC, GATTC, GAGTC and GACTC.

The restriction sequences for some of the most frequently used restriction endonucleases are listed in Table 4.1.

Table 4.1   The recognition sequences for some of the most frequently used restriction endonucleases

Enzyme	Organism	Recognition sequence[a]	Blunt or sticky end
*Eco*RI	*Escherichia coli*	GAATTC	sticky
*Bam*HI	*Bacillus amyloliquefaciens*	GGATCC	sticky
*Bgl*II	*Bacillus globigii*	AGATCT	sticky
*Pvu*I	*Proteus vulgaris*	CGATCG	sticky
*Pvu*II	*Proteus vulgaris*	CAGCTG	blunt
*Hin*dIII	*Haemophilus influenzae* R$_d$	AAGCTT	sticky
*Hin*fI	*Haemophilus influenzae* R$_f$	GANTC	sticky
*Sau*3A	*Staphylococcus aureus*	GATC	sticky
*Alu*I	*Arthrobacter luteus*	AGCT	blunt
*Taq*I	*Thermus aquaticus*	TCGA	sticky
*Hae*III	*Haemophilus aegyptius*	GGCC	blunt
*Not*I	*Nocardia otitidis-caviarum*	GCGGCCGC	sticky

[a] The sequence shown is that of one strand, given in the 5' to 3' direction. Note that almost all recognition sequences are palindromes: when both strands are considered they read the same in each direction, e.g.

```
            5' G A A T T C 3'
EcoRI:         | | | | | |
            3' C T T A A G 5'
```

4.2.3   Blunt ends and sticky ends

The exact nature of the cut produced by a restriction endonuclease is of considerable importance in the design of a gene cloning experiment. Many restriction endonucleases make a simple double-stranded cut in the middle of the recognition sequence (Figure 4.9(a)), resulting in a **blunt** or flush end. *Pvu*II and *Alu*I are examples of blunt-end cutters.

However, quite a large number of restriction endonucleases cut DNA in a slightly different way. With these enzymes the two DNA strands are not cut at exactly the same position. Instead the cleavage is staggered, usually by two or four nucleotides, so that the resulting DNA fragments have short single-stranded overhangs at each end (Figure 4.9(b)). These are called **sticky** or cohesive ends, as base pairing between them can stick the DNA molecule back together again (recall that sticky ends were encountered on p. 22 during the description of λ phage replication). One important feature of sticky end enzymes is that restriction endonucleases with different recognition sequences may produce the same sticky ends. *Bam*HI (recognition sequence GGATCC) and *Bgl*II (AGATCT) are examples – both produce GATC sticky ends (Figure 4.9(c)). The

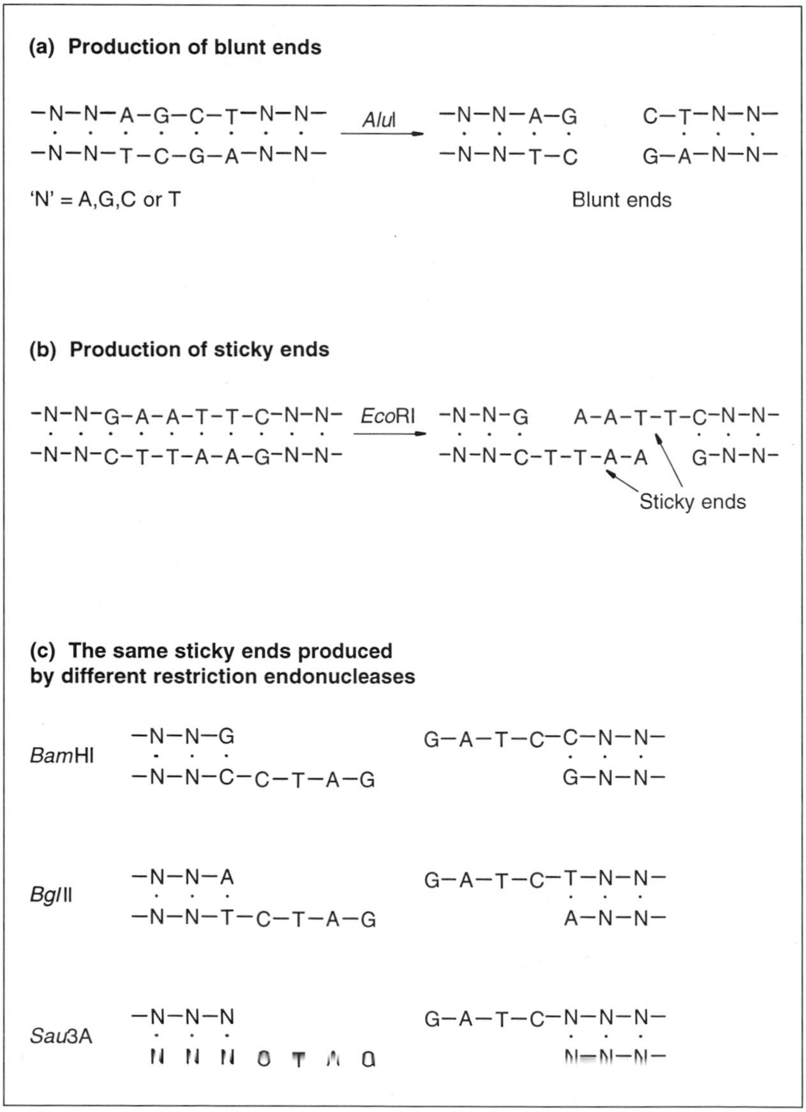

Figure 4.9 The ends produced by cleavage of DNA with different restriction endonucleases. (a) A blunt end produced by *Alu*I. (b) A sticky end produced by *Eco*RI. (c) The same sticky ends produced by *Bam*HI, *Bgl*II and *Sau*3A.

(a) **Production of blunt ends**

(b) **Production of sticky ends**

(c) **The same sticky ends produced by different restriction endonucleases** same sticky end is also produced by *Sau*3A, which recognizes only the tetranucleotide GATC. Fragments of DNA produced by cleavage with either of these enzymes can be joined to each other, as each fragment will carry a complementary sticky end.

4.2.4 The frequency of recognition sequences in a DNA molecule

The number of recognition sequences for a particular restriction endonuclease in a DNA molecule of known length can be calculated mathematically. A tetranucleotide sequence (e.g. GATC) should occur once every $4^4 = 256$ nucleotides, and a hexanucleotide (e.g. GGATCC) once every $4^6 = 4096$ nucleotides. These calculations assume that the nucleotides are ordered in a random fashion and that the four different nucleotides are present in equal proportions (i.e. the GC content = 50%). In practice, neither of these assumptions is entirely valid. For example, the λ DNA molecule, at 49 kb, should contain about 12 sites for a restriction endonuclease with a hexanucleotide recognition sequence. In fact these recognition sites occur less frequently (for example, six for *Bgl*II, five for *Bam*HI and only two for *Sal*I), a reflection of the fact that the GC content for λ is rather less than 50% (Figure 4.10(a)).

Furthermore, restriction sites are generally not evenly spaced out along a DNA molecule. If they were then digestion with a particular restriction endonuclease would give fragments of roughly equal sizes. Figure 4.10(b) shows the fragments produced by cutting λ DNA with *Bgl*II, *Bam*HI and *Sal*I. In each case there is a considerable spread of fragment sizes, indicating that in λ DNA the nucleotides are not randomly ordered.

The lesson to be learned from Figure 4.10 is that, although mathematics may give an idea of how many restriction sites to expect in a given DNA molecule, only experimental analysis can provide the true picture. We must therefore move on to consider how restriction endonucleases are used in the laboratory.

4.2.5 Performing a restriction digest in the laboratory

As an example we will consider how to digest a sample of λ DNA (concentration 125 μg/ml) with *Bgl*II.

First of all the required amount of DNA must be pipetted into a test-tube. The amount of DNA that will be restricted depends on the nature of the experiment; in this case we shall digest 2 μg of λ DNA, which is contained in 16 μl of the sample (Figure 4.11(a)). Very accurate micropipettes will therefore be needed.

Clearly the other main component in the reaction will be the restriction endonuclease, obtained from a commercial supplier as a pure solution of known concentration. But before adding the

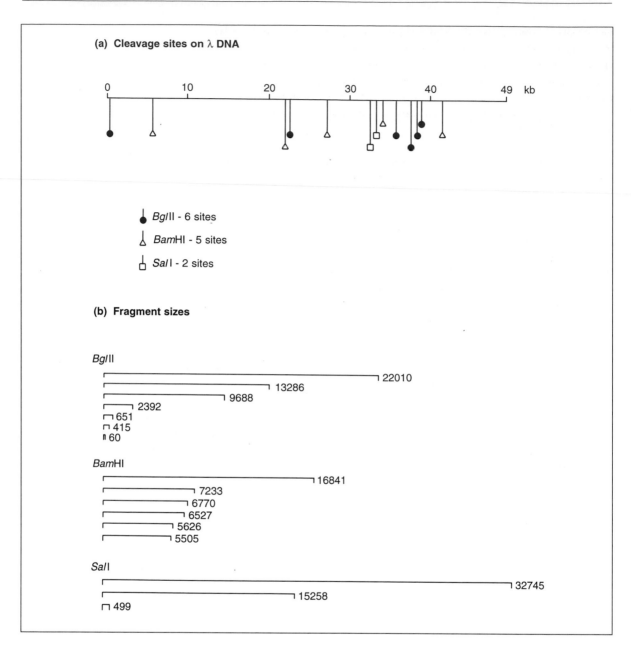

(a) Cleavage sites on λ DNA

*Bgl*II - 6 sites
*Bam*HI - 5 sites
*Sal*I - 2 sites

(b) Fragment sizes

*Bgl*II

22010
13286
9688
2392
651
415
60

*Bam*HI

16841
7233
6770
6527
5626
5505

*Sal*I

32745
15258
499

Figure 4.10 Restriction of the λ DNA molecule. (a) The positions of the recognition sequences for *Bgl*II, *Bam*HI and *Sal*I. (b) The fragments produced by cleavage with each of these restriction endonucleases; the numbers are the fragment sizes in base-pairs.

enzyme, the solution containing the DNA must be adjusted to provide the correct conditions to ensure maximal activity of the enzyme. Most restriction endonucleases function adequately at pH 7.4, but different enzymes vary in their requirements for ionic strength (provided usually by NaCl) and Mg^{2+} concentration (all Type II restriction endonucleases require Mg^{2+} in order to function). It is also advisable to add a reducing agent, such as dithiothreitol, which will stabilize the enzyme and prevent its inactivation. Providing the right conditions for the enzyme is very important – incorrect NaCl or Mg^{2+} concentrations may not only decrease the activity of the restriction endonuclease, but may also cause changes in the specificity of the enzyme, so that DNA cleavage occurs at additional, non-standard recognition sequences.

The composition of a suitable buffer for *Bgl*II is shown in Table 4.2. This buffer is ten times the working concentration, and is diluted by being added to the reaction mixture. In our example, a suitable final volume for the reaction mixture would be 20 μl, so we add 2 μl of $10 \times$ *Bgl*II buffer to the 16 μl of DNA already present (Figure 4.11(b)).

The restriction endonuclease can now be added. By convention, 1 unit of enzyme is defined as the quantity needed to cut 1 μg of DNA in 1 hour, so we need 2 units of *Bgl*II to cut 2 μg of λ DNA. *Bgl*II is frequently obtained at a concentration of 4 units/μl, so 0.5 μl will be sufficient to cleave the DNA. The final ingredients in the reaction mixture are therefore 0.5 μl *Bgl*II + 1.5 μl water, giving a final volume of 20 μl (Figure 4.11(c)).

The last factor to consider is incubation temperature. Most

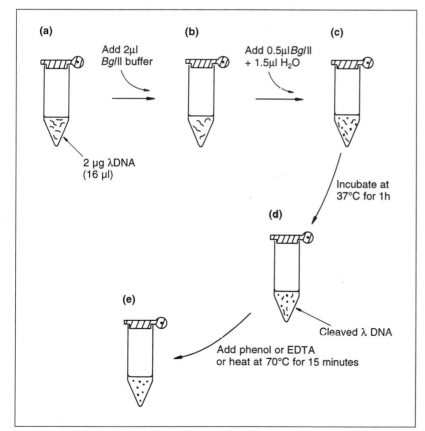

Figure 4.11 Performing a restriction digest in the laboratory. See the text for details.

Table 4.2 A 10 × buffer suitable for restriction of DNA with *Bgl*II

Component	Concentration (mM)
Tris–HCl, pH7.4	500
MgCl$_2$	100
NaCl	500
Dithiothreitol	10

restriction endonucleases, including *Bgl*II, work best at 37°C, but a few have different requirements. *Taq*I, for example, is purified from a bacterium called *Thermus aquaticus*, which normally lives at high temperatures in habitats such as hot springs. Restriction digests with *Taq*I must be incubated at 65°C to obtain maximum enzyme activity.

After 1 hour the restriction should be complete (Figure 4.11(d)). If the DNA fragments produced by restriction are to be used in cloning experiments, then the enzyme must somehow be destroyed so that it does not accidentally digest other DNA molecules that may be added at a later stage. There are several ways of 'killing' the enzyme. For many a short incubation at 70°C is sufficient, for others phenol extraction or the addition of EDTA (which binds Mg^{2+} ions preventing restriction endonuclease action) is used (Figure 4.11(e)).

4.2.6 Analysing the result of restriction endonuclease cleavage

A restriction digest will result in a number of DNA fragments, the sizes of which depend on the exact positions of the recognition sequences for the endonuclease in the original molecule (Figure 4.10). Clearly a way of determining the number and sizes of the fragments is needed if restriction endonucleases are to be of use in gene cloning. Whether or not a DNA molecule is cut at all can be determined fairly easily by testing the viscosity of the solution. Larger DNA molecules result in a more viscous solution than smaller ones, so cleavage is associated with a decrease in viscosity. However, working out the number and sizes of the individual cleavage products is more difficult. In fact, for several years this was one of the most tedious aspects of experiments involving DNA. Eventually the problems were solved in the early 1970s when the technique of gel electrophoresis was developed

(a) Separation of molecules by gel electrophoresis   DNA molecules, like proteins and many other biological compounds, carry an electric charge, negative in the case of DNA. Consequently, when DNA molecules are placed in an electric field they migrate towards the positive pole (Figure 4.12(a)). The rate of migration of a molecule depends on two factors, its shape and its charge-to-mass ratio. Unfortunately, most DNA molecules are the same shape and all have very similar charge-to-mass ratios. Fragments of different sizes cannot therefore be separated by standard **electrophoresis**.

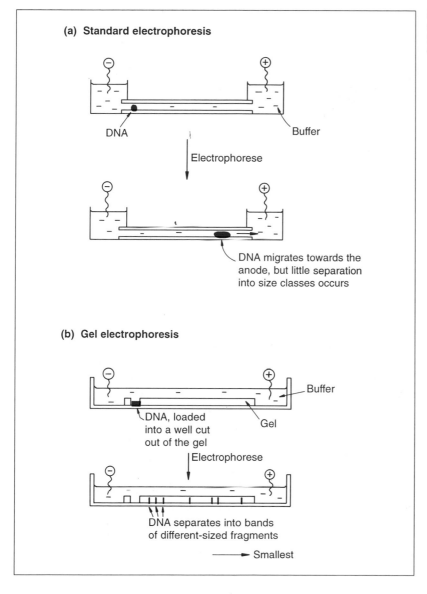

Figure 4.12 Standard electrophoresis (a) does not separate DNA fragments of different sizes, whereas gel electrophoresis (b) does.

However, the size of the DNA molecule does become a factor if the electrophoresis is performed in a gel. A gel, which is usually made of agarose, polyacrylamide or a mixture of the two, comprises a complex network of pores, through which the DNA molecules must travel to reach the positive electrode. The smaller the DNA molecule, the faster it can migrate through the gel. **Gel electrophoresis** therefore separates DNA molecules according to their size (Figure 4.12(b)).

In practice the composition of the gel determines the sizes of the DNA molecules that can be separated. A 0.5-cm thick slab of 0.5% agarose, which has relatively large pores, would be used for molecules in the size range 1 to 30 kb, allowing, for example, molecules of 10 and 12 kb to be clearly distinguished. At the other end of the scale, a very thin (0.3-mm) 40% polyacrylamide gel, with extremely small pores, would be used to separate much smaller DNA molecules, in the range 1 to 300 bp, and could distinguish molecules differing in length by just a single nucleotide.

(b) Visualizing DNA molecules in a gel

(i) Staining The easiest way to see the results of a gel electrophoresis experiment is to stain the gel with a compound that will make the DNA visible. Ethidium bromide, already described on p. 42 as a means of visualizing DNA in CsCl gradients, is also routinely used to stain DNA in agarose and polyacrylamide gels (Figure 4.13). Bands showing the positions of the different size classes of DNA fragment are clearly visible under ultraviolet irradiation after EtBr staining, so long as sufficient DNA is present.

(ii) Autoradiography of radioactively labelled DNA The one drawback with EtBr staining is that there is a limit to its sensitivity. If less than about 25 ng of DNA is present per band then it is unlikely that the results will show up with EtBr staining. For small amounts of DNA a much more sensitive detection method is needed.

Autoradiography provides an answer. If the DNA is **labelled**, before electrophoresis, by incorporation of a **radioactive marker** into the individual molecules, then the DNA can be visualized by placing an X-ray-sensitive photographic film over the gel. The radioactive DNA exposes the film, revealing the banding pattern (Figure 4.14).

A DNA molecule is usually labelled by incorporating nucleotides that carry a radioactive isotope of phosphorus, ^{32}P (Figure 4.15(a)). Several methods are available, the most popular being **nick translation** and **end-filling**.

Nick translation refers to the activity of DNA polymerase I (p. 57). Most purified samples of DNA contain some nicked molecules, however carefully the preparation has been carried out, so

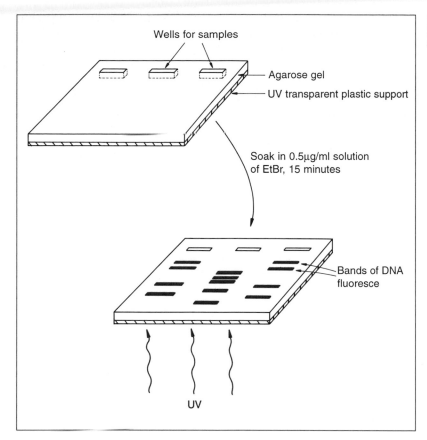

DNA polymerase I will be able to attach to the DNA and catalyse a strand replacement reaction (Figure 4.5(b)). This reaction requires a supply of nucleotides: if one of these is radioactively labelled, then the DNA molecule will itself become labelled (Figure 4.15(b)).

Nick translation can be used to label any DNA molecule but may under some circumstances also cause DNA cleavage. End-filling is a gentler method that rarely causes breakage of the DNA, but unfortunately can only be used to label DNA molecules that have sticky ends. The enzyme used is the Klenow fragment (p. 57), which 'fills in' a sticky end by synthesizing the complementary strand (Figure 4.15(c)). As with nick translation, if the end-filling reaction is carried out in the presence of labelled nucleotides, then the DNA itself becomes labelled.

Both nick translation and end-filling can label DNA to such an extent that very small quantities can be detected in gels by autoradiography. As little as 2 ng of DNA per band can be visualized under ideal conditions.

Figure 4.14 The use of autoradiography to visualize radioactively labelled DNA in an agarose gel.

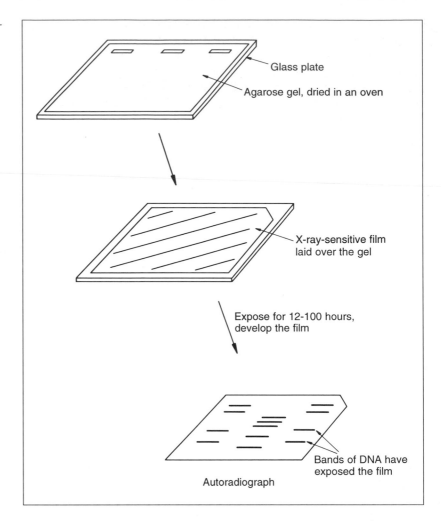

Glass plate

Agarose gel, dried in an oven

X-ray-sensitive film laid over the gel

Expose for 12–100 hours, develop the film

Bands of DNA have exposed the film

Autoradiograph

4.2.7 Estimation of the sizes of DNA molecules

Gel electrophoresis will separate different-sized DNA molecules, with the smallest molecules travelling the greatest distance towards the positive electrode. If several DNA fragments of varying sizes are present (the result of a successful restriction digest, for example), then a series of bands will appear in the gel. How can the sizes of these fragments be determined?

The most accurate method is to make use of the mathematical relationship that links migration rate to molecular weight. The relevant formula is

$$D = a - b(\log M)$$

where D is the distance moved, M is the molecular weight, and a and b are constants that depend on the electrophoresis conditions.

However, a much simpler though less precise way of estimating DNA fragment sizes is generally used. A standard restriction digest, comprising fragments of known size, is usually included in each electrophoresis gel that is run. Restriction digests of λ DNA are often used in this way as size markers. For example, *Hind*III cleaves λ DNA into eight fragments, ranging in size from 125 bp for the smallest to over 23 kb for the largest. As the sizes of the fragments in this digest are known, the fragment sizes in the experimental digest can be estimated by comparing the positions of the bands in the two tracks (Figure 4.16). Although not precise, this method can be performed with as little as a 5% error, which is quite satisfactory for most purposes.

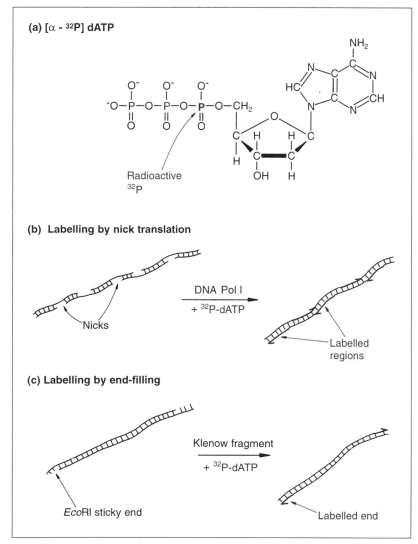

(a) [α - ³²P] dATP

Radioactive
³²P

(b) Labelling by nick translation

DNA Pol I
+ ³²P-dATP

Nicks

Labelled regions

(c) Labelling by end-filling

Klenow fragment
+ ³²P-dATP

*Eco*RI sticky end

Labelled end

Figure 4.15 Radioactive labelling. (a) The structure of [α − ³²P]dATP. (b) Labelling DNA by nick translation. (c) Labelling DNA by end-filling.

Figure 4.16 Estimation of the sizes of DNA fragments in an agarose gel. (a) A rough estimate of fragment size can be obtained by eye. (b) A more accurate measurement of fragment size is gained by using the mobilities of the *Hind*III-fragments to construct a calibration curve; the sizes of the unknown fragments can then be determined from the distances they have migrated.

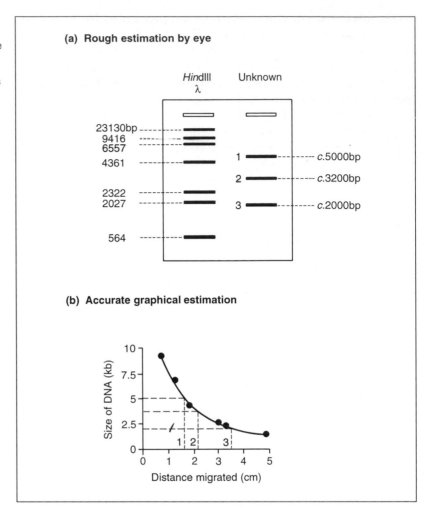

4.2.8 Mapping the positions of different restriction sites on a DNA molecule

So far we have considered how to determine the number and sizes of the DNA fragments produced by restriction endonuclease cleavage. The next step in **restriction analysis** is to construct a map showing the relative positions on the DNA molecule of the recognition sequences for a number of different enzymes. Only when a **restriction map** is available can the correct restriction endonucleases be selected for the particular cutting manipulation that is required (Figure 4.17).

To construct a restriction map, a series of restriction digests must be performed. Firstly, the number and sizes of the fragments produced by each restriction endonuclease must be determined by gel electrophoresis, followed by comparison with size markers (Figure

4 18). This information must then be supplemented by a series of **double digestions**, in which the DNA is cut by two restriction endonucleases at once. It may be possible to perform a double digestion in one step, if both enzymes have similar requirements for pH, Mg^{2+} concentration, etc. Alternatively, the two digestions may have to be carried out one after the other, adjusting the reaction mixture after the first digestion to provide a different set of conditions for the second enzyme.

Comparing the results of single and double digests will allow many, if not all, of the restriction sites to be mapped (Figure 4.18). Ambiguities can usually be resolved by **partial digestion**, carried out under conditions that result in cleavage of only a limited number of the restriction sites on any DNA molecule. Partial digestion is achieved normally either by a short incubation period, so the enzyme does not have time to cut all the restriction sites, or by incubation at a low temperature (e.g. 4°C rather than 37°C), which limits the activity of the enzyme. The result of a partial digestion is a complex pattern of bands in an electrophoresis gel. As well as the standard fragments, produced by total digestion, additional sizes are seen. These are molecules that comprise two adjacent restriction fragments, separated by a site that has

Figure 4.17 Using a restriction map to work out which restriction endonucleases should be used to obtain DNA fragments containing individual genes.

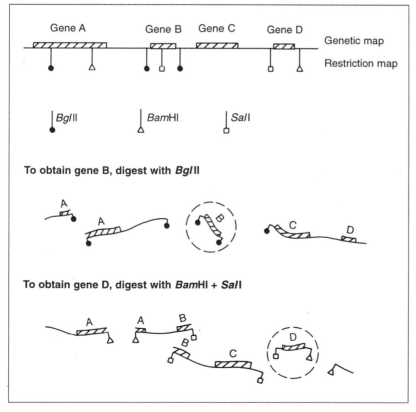

not been cleaved. Their sizes indicate which restriction fragments in the complete digest are next to one another in the uncut molecule (Figure 4.18).

Figure 4.18 Restriction mapping. This example shows how the positions of the *Xba*I, *Xho*I and *Kpn*I sites on the λ DNA molecule can be determined.

Single and double digestions

Enzyme	Number of fragments	Sizes (kb)
*Xba*I	2	24.0 24.5
*Xho*I	2	15.0 33.5
*Kpn*I	3	1.5 17.0 30.0
*Xba*I + *Xho*I	3	9.0 15.0 24.5
*Xba*I + *Kpn*I	4	1.5 6.0 17.0 24.0

Conclusions:

1. As λ DNA is linear, the number of restriction sites for each enzyme is *Xba*i 1, *Xho*I 1, *Kpn*I 2.

2. The *Xba*I and *Xho*I sites can be mapped:

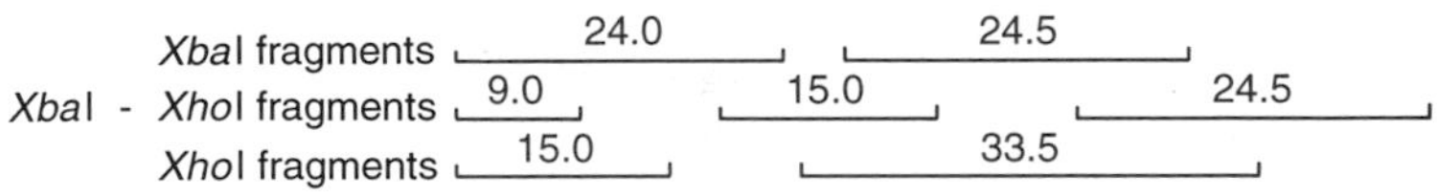

*Xba*I fragments	24.0 ┆ 24.5
*Xba*I - *Xho*I fragments	9.0 ┆ 15.0 ┆ 24.5
*Xho*I fragments	15.0 ┆ 33.5

The only possibility is:

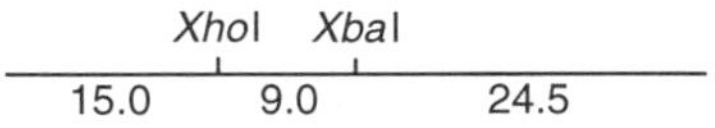

```
          Xhol   Xbal
        ─────┴────┴──────────
        15.0   9.0    24.5
```

3. All the *Kpn*I sites fall in the 24.5 kb *Xba*I fragment, as the 24.0 kb fragment is intact after *Xba*I - *Kpn*I double digestion. The order of the *Kpn*I fragments can be determined only by partial digestion.

Partial digestion

Enzyme	Fragment sizes (kb)
*Kpn*I - limiting conditions	1.5, 17.0, 18.5, 30.0, 31.5, 48.5

Conclusions:

48.5 kb fragment = uncut λ.
1.5, 17.0 and 30.0 kb fragments are products of complete digestion.
18.5 and 31.5 kb fragments are products of partial digestion.

The *Kpn*I map must be:

```
                          Kpnls
        ──────────────────┴┴──────
              30.0      1.5  17.0
```

Therefore the complete map is:

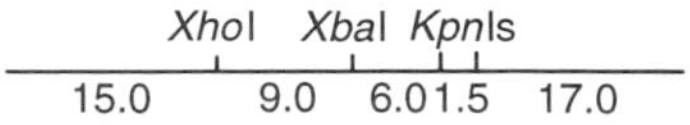

```
            Xhol   Xbal Kpnls
        ─────┴──────┴───┴┴──────
           15.0   9.0  6.01.5  17.0
```

4.3 LIGATION – JOINING DNA MOLECULES TOGETHER

The final step in construction of a recombinant DNA molecule is the joining together of the vector molecule and the DNA to be cloned (Figure 4.19). This process is referred to as ligation, and the enzyme that catalyses the reaction is called DNA ligase.

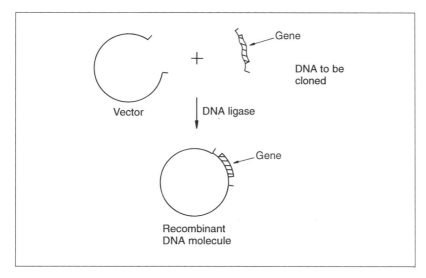

Figure 4.19 Ligation: the final step in construction of a recombinant DNA molecule.

4.3.1 The mode of action of DNA ligase

All living cells produce DNA ligases, but the enzyme used in genetic engineering is usually purified from *E. coli* bacteria that have been infected with T4 phage. Within the cell the enzyme carries out the very important function of repairing any discontinuities that may arise in one of the strands of a double-stranded molecule (Figure 4.4(a)). A discontinuity is quite simply a position where a phosphodiester bond between adjacent nucleotides is missing (contrast this with a 'nick', where one or more nucleotides are absent). Although discontinuities may arise by chance breakage of the cell's DNA molecules, they are also a natural result of processes such as DNA replication and recombination. Ligases therefore play several vital roles in the cell.

In the test-tube, purified DNA ligases, as well as repairing single-strand discontinuities, will also join together individual DNA molecules or the two ends of the same molecule. The chemical reaction involved in ligating two molecules is exactly the same as discontinuity repair, except that two phosphodiester bonds must be made, one for each stand (Figure 4.20(a)).

Figure 4.20 The different joining reactions catalysed by DNA ligase. (a) Ligation of blunt-ended molecules. (b) Ligation of sticky-ended molecules.

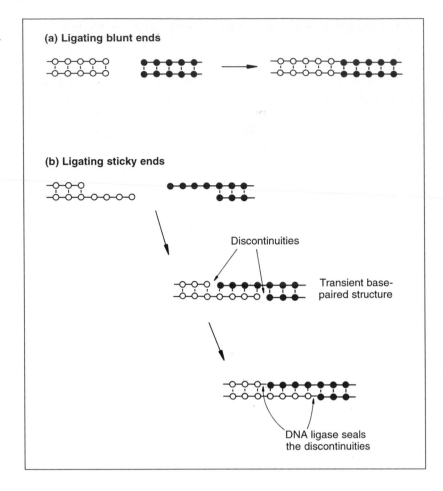

4.3.2 Sticky ends increase the efficiency of ligation

The ligation reaction in Figure 4.20(a) shows two blunt-ended fragments being joined together. Although this reaction can be carried out in the test-tube, it is not very efficient. This is because the ligase is unable to 'catch hold' of the molecule to be ligated, and has to wait for chance associations to bring the ends together. If possible, blunt-end ligation should be performed at high DNA concentrations, to increase the chances of the ends of the molecules coming together in the correct way.

In contrast, ligation of complementary sticky ends is much more efficient. This is because compatible sticky ends can base-pair with one another by hydrogen bonding (Figure 4.20(b)) forming a relatively stable structure for the enzyme to work on. If the phosphodiester bonds are not synthesized fairly quickly, then the sticky ends will fall apart again. These transient, base-paired structures do, however, increase the efficiency of ligation by increasing the length of time the ends are in contact with one another.

4.3.3 Putting sticky ends on to a blunt-ended molecule

For the reasons detailed in the preceding section, compatible sticky ends are desirable on the DNA molecules to be ligated together in a gene cloning experiment. Often these sticky ends can be provided by digesting both the vector and the DNA to be cloned with the same restriction endonuclease, or with different enzymes that produce the same sticky end. However, it is not always possible to do this. A common situation is where the vector molecule has sticky ends, but the DNA fragments to be cloned are blunt-ended. Under these circumstances one of three methods can be used to put the correct sticky ends on to the DNA fragments.

(a) Linkers   The first of these methods involves the use of **linkers**. These are short pieces of double-stranded DNA, of known nucleotide sequence, that are synthesized in the test-tube. A typical linker is shown in Figure 4.21(a). It is blunt-ended, but contains a restriction site, *Bam*HI in the example shown. DNA ligase will attach linkers to the ends of larger blunt-ended DNA molecules. Although a blunt-end ligation, this particular reaction can be

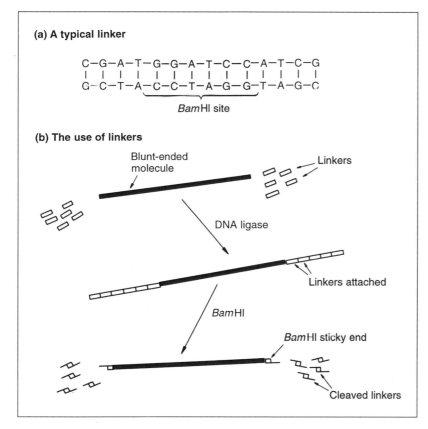

(a) A typical linker

*Bam*HI site

(b) The use of linkers

Blunt-ended molecule

Linkers

DNA ligase

Linkers attached

*Bam*HI

*Bam*HI sticky end

Cleaved linkers

Figure 4.21 Linkers and their use. (a) The structure of a typical linker. (b) The attachment of linkers to a blunt-ended molecule.

performed very efficiently because synthetic oligonucleotides, such as linkers, can be made in very large amounts and added into the ligation mixture at a high concentration.

More than one linker will attach to each end of the DNA molecule, producing the chain structure shown in Figure 4.21(b). However, digestion with *Bam*HI cleaves the chains at the recognition sequences, producing a large number of cleaved linkers and the original DNA fragment, now carrying *Bam*HI sticky ends. This modified fragment is ready for ligation into a cloning vector restricted with *Bam*HI.

(b) Adaptors There is one potential drawback with the use of linkers. Consider what would happen if the blunt-ended molecule shown in Figure 4.21(b) contained one or more *Bam*HI recognition sequences. If this was the case then the restriction step needed to cleave the linkers and produce the sticky ends would also cleave the blunt-ended molecule (Figure 4.22). The resulting fragments will have the correct sticky ends, but that is no consolation if the gene contained in the blunt-ended fragment has now been broken into pieces.

The second method of attaching sticky ends to a blunt-ended molecule is designed to avoid this problem. **Adaptors**, like linkers, are short synthetic oligonucleotides. However, unlike linkers, an adaptor is synthesized so that it already has one sticky end (Figure

Figure 4.22 A possible problem with the use of linkers. Compare this situation with the desired result of *Bam*HI restriction, as shown in Figure 4.21(b)).

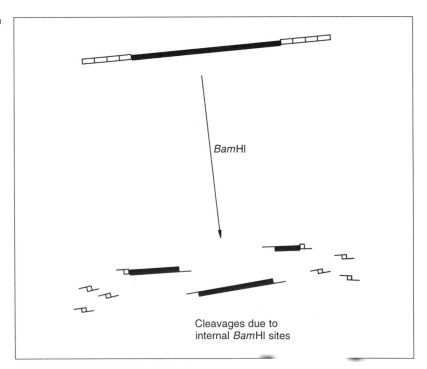

BamHI

Cleavages due to internal *Bam*HI sites

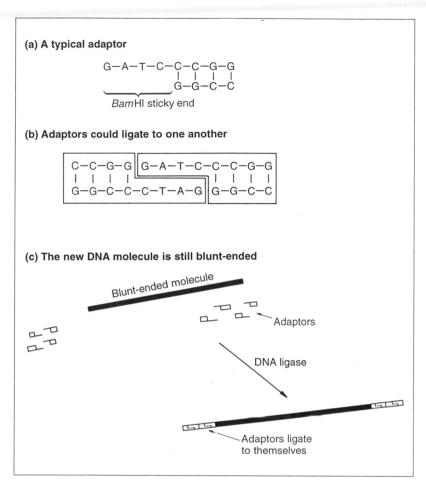

(a) A typical adaptor

$$G-A-T-C-C-C-G-G$$
$$\qquad\qquad\ |\ \ |\ \ |\ \ |$$
$$\qquad\qquad G-G-C-C$$

BamHI sticky end

(b) Adaptors could ligate to one another

$$C-C-G-G\ |\ G-A-T-C-C-C-G-G$$
$$|\ \ |\ \ |\ \ |\qquad\qquad\quad\ |\ \ |\ \ |\ \ |$$
$$G-G-C-C-C-T-A-G\ |\ G-G-C-C$$

(c) The new DNA molecule is still blunt-ended

Blunt-ended molecule

Adaptors

DNA ligase

Adaptors ligate
to themselves

Figure 4.23 Adaptors and the potential problem with their use. (a) A typical adaptor. (b) Two adaptors could ligate to one another to produce a molecule similar to a linker, so that (c) after ligation of adaptors a blunt-ended molecule is still blunt-ended and the restriction step is still needed.

4.23(a)). The idea is of course to ligate the blunt end of the adaptor to the blunt ends of the DNA fragment, to produce a new molecule with sticky ends. This may appear to be a simple method but in practice a new problem arises. The sticky ends of individual adaptor molecules could of course base-pair with each other to form dimers (Figure 4.23(b)), so that the new DNA molecule is still blunt-ended (Figure 4.23(c)). The sticky ends could be recreated by digestion with a restriction endonuclease, but that would defeat the purpose of using adaptors in the first place.

The answer to the problem lies in the precise chemical structure of the ends of the adaptor molecule. Normally, the two ends of a polynucleotide strand are chemically distinct, a fact that will be clear from a careful examination of the polymeric structure (Figure 4.24(a)). One end, referred to as the 5'-terminus, carries a phosphate group (5'-P); the other, the 3'-terminus, has a hydroxyl group (3'-OH). In the double helix the two strands are antiparallel (Figure 4.24(b)), so each end of a double-stranded molecule

Figure 4.24 The distinction between the 5'- and 3'-termini of a polynucleotide.

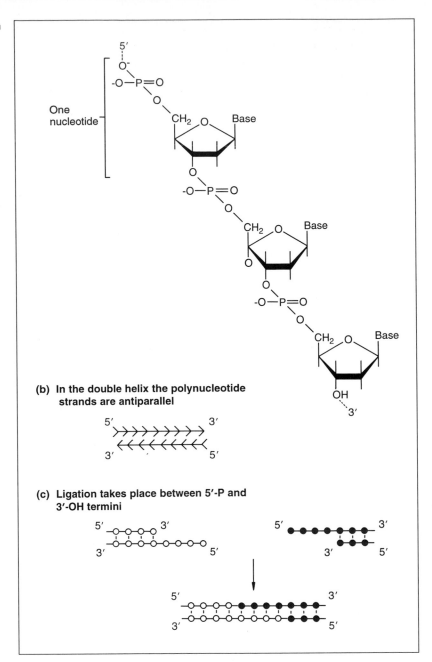

(b) **In the double helix the polynucleotide strands are antiparallel**

(c) **Ligation takes place between 5'-P and 3'-OH termini**

consists of one 5'-P terminus and one 3'-OH terminus. Ligation normally takes place between the 5'-P and 3'-OH ends (Figure 4.24(c)).

Adaptor molecules are synthesized so that the blunt end is the same as 'natural' DNA, but the sticky end is different. The 3'-OH terminus of the sticky end is the same as usual, but the 5'-P terminus is modified; it lacks the phosphate group, and is in fact a 5'-OH terminus (Figure 4.25(a)). DNA ligase is unable to form a phosphodiester bridge between 5'-OH and 3'-OH ends. The result is that, although base-pairing is always occurring between the sticky ends of adaptor molecules, the association is never stabilized by ligation (Figure 4.25(b)).

Adaptors can therefore be ligated to a DNA molecule but not to themselves. After the adaptors have been attached the abnormal 5'-OH terminus is converted to the natural 5'-P form by treatment with the enzyme polynucleotide kinase (p. 59), producing a sticky-ended fragment that can be inserted into an appropriate vector.

(c) Producing sticky ends by homopolymer tailing   The technique of **homopolymer tailing** offers a radically different approach to the production of sticky ends on a blunt-ended DNA molecule. A homopolymer is simply a polymer in which all the subunits are

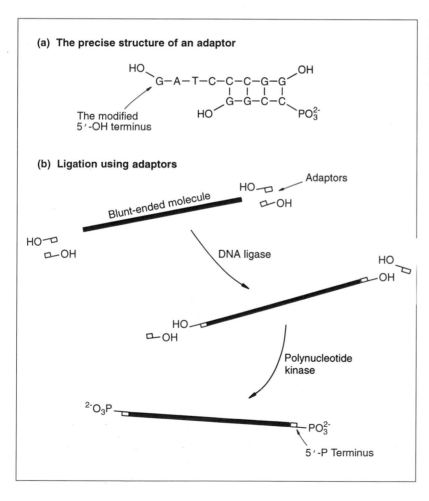

Figure 4.25 The use of adaptors. (a) The actual structure of an adaptor, showing the modified 5'-OH terminus. (b) Conversion of blunt ends to sticky ends through the attachment of adaptors.

(a) The precise structure of an adaptor

The modified
5'-OH terminus

(b) Ligation using adaptors

Blunt-ended molecule

Adaptors

DNA ligase

Polynucleotide
kinase

5'-P Terminus

Figure 4.26 Homopolymer tailing. (a) Synthesis of a homopolymer tail. (b) Construction of a recombinant DNA molecule from a tailed vector plus tailed insert DNA. (c) Repair of the recombinant DNA molecule.

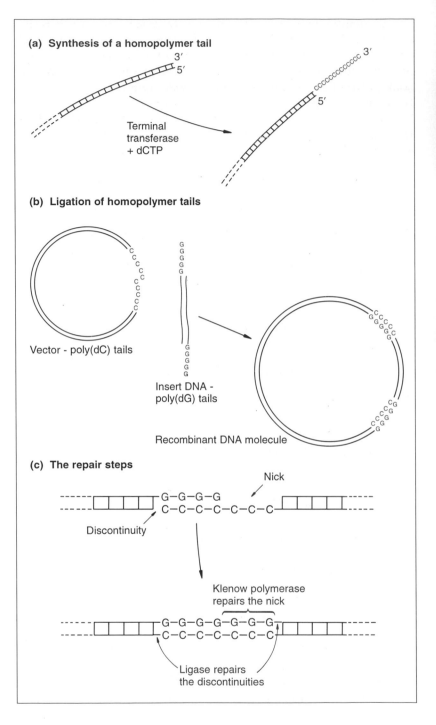

the same. A DNA strand made up entirely of, say, deoxyguanosine is an example of a homopolymer, and is referred to as poly-deoxyguanosine or poly(dG).

Tailing involves using the enzyme terminal deoxynucleotidyl transferase (p. 59) to add a series of nucleotides on to the 3′-OH termini of a double-stranded DNA molecule. If this reaction is carried out in the presence of just one deoxynucleotide, then a homopolymer tail will be produced (Figure 4.26(a)).

Of course, to be able to ligate together two tailed molecules, the homopolymers must be complementary. Frequently poly(dC) tails are attached to the vector and poly(dG) to the DNA to be cloned. Base-pairing between the two occurs when the DNA molecules are mixed (Figure 4.26(b)).

In practice the poly(dG) and poly(dC) tails are not usually exactly the same length, and base-paired recombinant molecules will have nicks as well as discontinuities (Figure 4.26(c)). Repair is therefore a two-step process, using Klenow polymerase to fill in the nicks followed by DNA ligase to synthesize the final phosphodiester bonds. This repair reaction does not always have to be performed in the test-tube. If the complementary homopolymer tails are longer than about 20 nucleotides, then quite stable base-paired associations will be formed. A recombinant DNA molecule, held together by base-pairing although not completely ligated, will often be stable enough to be introduced into the host cell in the next stage of the cloning experiment (Figure 1.1). Once inside the host, the cell's own DNA polymerase and DNA ligase will repair the recombinant DNA molecule, completing the construction begun in the test-tube.

FURTHER READING

Jacobsen, H., Klenow, H. and Overgaard-Hansen, K. (1974) The N-terminal amino-acid sequences of DNA polymerase I from *Escherichia coli* and of the large and small fragments obtained by a limited proteolysis. *European Journal of Biochemistry*, **45**, 623–7 – production of the Klenow fragment of DNA polymerase I.

Smith, H. O. and Wilcox, K. W. (1970) A restriction enzyme from *Haemophilus influenzae*. *Journal of Molecular Biology*, **51**, 379–91 – one of the first full descriptions of a restriction endonuclease.

Roberts, R. J. and Macelis, D. (1993) Restriction enzymes and methylases. *Nucleic Acids Research*, **21**, 3125–37 – a comprehensive list of all the known restriction endonucleases and their recognition sequences.

McDonell, M. W., Simon, M. N. and Studier, F. W. (1977) Analysis of restriction fragments of T7 DNA and determination of molecular weights by electrophoresis in neutral and alkaline gels. *Journal of Molecular Biology*, **110**, 119–46 – an early example of the use of agarose gel electrophoresis in the analysis of restriction fragment sizes.

Lobban, P. and Kaiser, A. D. (1973) Enzymatic end-to-end joining of DNA molecules. *Journal of Molecular Biology*, **79**, 453–71 – ligation.

Rothstein, R. J. *et al.* (1979) Synthetic adaptors for cloning DNA. *Methods in Enzymology*, **68**, 98–109.

Introduction of DNA into living cells   5

The manipulations described in Chapter 4 allow the molecular biologist to create novel recombinant DNA molecules. The next step in a gene cloning experiment is to introduce these molecules into living cells, usually bacteria, which will then grow and divide to produce clones (Figure 1.1). Strictly speaking, the word 'cloning' refers only to the later stages of the procedure, and not to the construction of the recombinant DNA molecule itself.

Cloning serves two main purposes. Firstly, it allows a large number of recombinant DNA molecules to be produced from a limited amount of starting material. At the outset only a few nanograms of recombinant DNA may be available, but each bacterium that takes up a plasmid will divide numerous times to produce a colony, each cell of which contains multiple copies of the molecule. Several micrograms of recombinant DNA can usually be prepared from a single bacterial colony, representing a thousand-fold increase over the starting amount (Figure 5.1). If the colony is used not as a source of DNA, but as an inoculum for a liquid culture, then the resulting cells may provide milligrams of DNA, a millionfold increase in yield. In this way cloning can supply the large amounts of DNA needed for molecular biological studies of gene structure and expression (Chapters 9 and 10).

The second important function of cloning can be described as purification. The manipulations that result in a recombinant DNA molecule can only rarely be controlled to the extent that no other DNA molecules are present at the end of the procedure. The ligation mixture may contain, in addition to the desired recombinant molecule, any number of the following (Figure 5.2(a)):

1. unligated vector molecules;
2. unligated DNA fragments;
3. vector molecules that have recircularized without new DNA being inserted ('self-ligated' vector);
4. recombinant DNA molecules that carry the wrong inserted DNA fragment.

Figure 5.1 Cloning can supply large amounts of recombinant DNA.

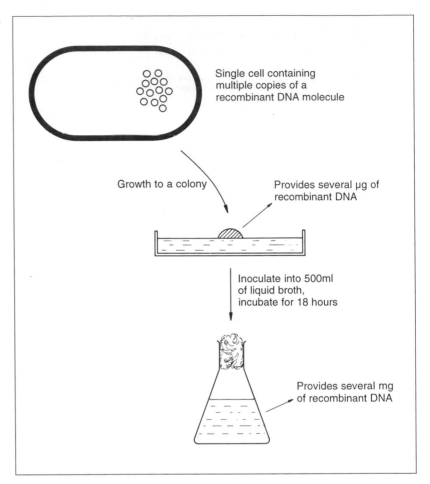

Single cell containing multiple copies of a recombinant DNA molecule

Growth to a colony

Provides several μg of recombinant DNA

Inoculate into 500ml of liquid broth, incubate for 18 hours

Provides several mg of recombinant DNA

Unligated molecules rarely cause a problem because, even though they may be taken up by bacterial cells, only under exceptional circumstances will they be replicated. It is much more likely that the host enzymes will degrade these pieces of DNA. On the other hand, self-ligated vector molecules and incorrect recombinant plasmids will be replicated just as efficiently as the desired molecule (Figure 5.2(b)). However, purification of the desired molecule can still be achieved through cloning because it is extremely unusual for any one cell to take up more than one DNA molecule. Each cell gives rise to a single colony, so each of the resulting clones consists of cells that all contain the same molecule. Of course different colonies contain different molecules: some contain the desired recombinant DNA molecule, some have different recombinant molecules, and some self-ligated vector. The problem therefore becomes a question of identifying the colonies that contain the correct recombinant plasmids.

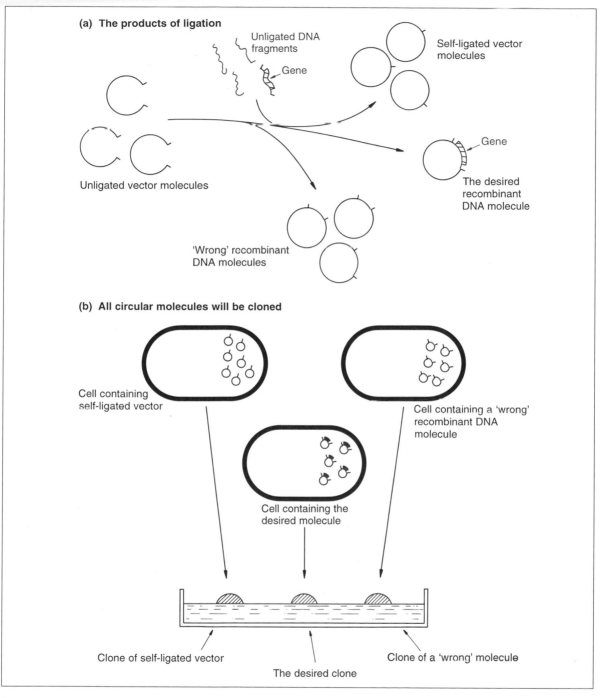

(a) The products of ligation

Unligated DNA fragments

Gene

Self-ligated vector molecules

Unligated vector molecules

Gene

The desired recombinant DNA molecule

'Wrong' recombinant DNA molecules

(b) All circular molecules will be cloned

Cell containing self-ligated vector

Cell containing a 'wrong' recombinant DNA molecule

Cell containing the desired molecule

Clone of self-ligated vector

The desired clone

Clone of a 'wrong' molecule

Figure 5.2 Cloning is analogous to purification. From a mixture of different molecules, clones containing copies of just one molecule can be obtained.

This chapter is concerned with the way in which plasmid and phage vectors, and recombinant molecules derived from them, are introduced into bacterial cells. During the course of the chapter it will become apparent that selection for colonies containing recombinant molecules, as opposed to colonies containing self-ligated vector, is relatively easy. The more difficult proposition of how to distinguish clones containing the correct recombinant DNA molecule from all the other recombinant clones will be tackled in Chapter 8.

5.1 TRANSFORMATION – THE UPTAKE OF DNA BY BACTERIAL CELLS

Most species of bacteria are able to take up DNA molecules from the medium in which they grow. Often a DNA molecule taken up in this way will be degraded, but occasionally it will be able to survive and replicate in the host cell. In particular this will happen if the DNA molecule is a plasmid with an origin of replication recognized by the host.

Uptake and stable retention of a plasmid is usually detected by looking for expression of the genes carried by the plasmid (p. 13). For example, *E. coli* cells are normally sensitive to the growth inhibitory effects of the antibiotics ampicillin and tetracycline. However, cells that contain the plasmid pBR322 (p. 108), one of the first cloning vectors to be developed back in the 1970s, are resistant to these antibiotics. This is because pBR322 carries two sets of genes, one gene that codes for a β-lactamase enzyme that modifies ampicillin into a form that is non-toxic to the bacterium, and a second set of genes that code for enzymes that detoxify tetracycline. Uptake of pBR322 can be detected because the *E. coli* cells are **transformed** from ampicillin- and tetracycline-sensitive (amp^{S}tet^S) to ampicillin- and tetracycline-resistant (amp^{R}tet^R).

In recent years the term **transformation** has been extended to include uptake of any DNA molecule by any type of cell, regardless of whether the uptake results in a detectable change in the cell, or whether the cell involved is bacterial, fungal, animal or plant.

5.1.1 Not all species of bacteria are equally efficient at DNA uptake

In nature, transformation is probably not a major process by which bacteria obtain genetic information. This is reflected by the fact that in the laboratory only a few species (notably members of the genera *Bacillus* and *Streptococcus*) can be transformed with ease. Close

study of these organisms has revealed that they possess sophisticated mechanisms for DNA binding and uptake.

Most species of bacteria, including *E. coli*, take up only limited
amounts of DNA under normal circumstances. In order to transform these species efficiently, the bacteria have to undergo some
form of physical and/or chemical treatment that enhances their
ability to take up DNA. Cells that have undergone this treatment
are said to be **competent**.

5.1.2 Preparation of competent *E. coli* cells

As with many breakthroughs in recombinant DNA technology, the
key development as far as transformation was concerned occurred
in the early 1970s, when it was observed that *E. coli* cells that had
been soaked in an ice-cold salt solution were more efficient at DNA
uptake than unsoaked cells. A solution of 50 mM calcium chloride
is traditionally used, although other salts, notably rubidium chloride, are also effective.

Exactly why this treatment works is not understood. Possibly
$CaCl_2$ causes the DNA to precipitate on to the outside of the cells,
or perhaps the salt is responsible for some kind of change in the
cell wall that improves DNA binding. In any case, soaking in $CaCl_2$
affects only DNA binding, and not the actual uptake into the cell.
When DNA is added to treated cells, it remains attached to the cell
exterior, and is not at this stage transported into the cytoplasm
(Figure 5.3). The actual movement of DNA into competent cells is
stimulated by briefly raising the temperature to 42°C. Once again,
the exact reason why this heat-shock is effective is not understood.

5.1.3 Selection for transformed cells

Transformation of competent cells is an inefficient procedure, however carefully the cells have been prepared. Although 1 ng of pUC8
can yield 1000–10 000 transformants, this represents the uptake of
only 0.01% of all the available molecules. Furthermore, 10 000
transformants is only a very small proportion of the total number
of cells that are present in a competent culture. This last fact means
that some way must be found to distinguish a cell that has taken
up a plasmid from the many thousands that have not been transformed.

The answer is to make use of a **selectable marker** carried by the
plasmid. A selectable marker is simply a gene that provides a
transformed cell with a new characteristic, one that is not possessed by a non-transformant. A good example of a selectable
marker is the ampicillin resistance gene of pBR322. After a transformation experiment with pBR322, only those *E. coli* cells that
have taken up a plasmid will be $amp^R tet^R$ and able to form

Figure 5.3 The binding and uptake of DNA by a competent bacterial cell.

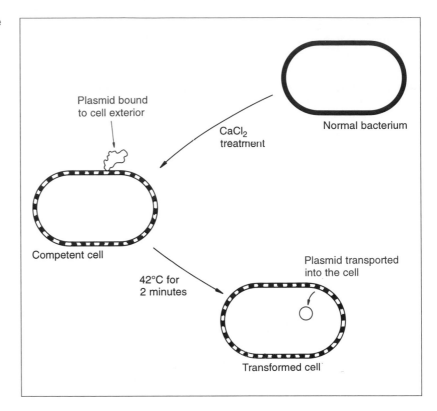

colonies on an agar medium that contains ampicillin or tetracycline (Figure 5.4); non-transformants, which are still amp^{S}tet^S, will not produce colonies on the selective medium. Transformants and non-transformants are therefore easily distinguished.

Most plasmid cloning vectors carry at least one gene that confers antibiotic resistance on the host cells, with selection of transformants being achieved by plating on to an agar medium that contains the relevant antibiotic. Bear in mind, however, that resistance to the antibiotic is not due merely to the presence of the plasmid in the transformed cells. The resistance gene on the plasmid must also be expressed, so that the enzyme that detoxifies the antibiotic is synthesized. Expression of the resistance gene begins immediately after transformation, but it will be a few minutes before the cell contains enough of the enzyme to be able to withstand the toxic effects of the antibiotic. For this reason the transformed bacteria should not be plated on to the selective medium immediately after the heat-shock treatment, but first placed in a small volume of liquid medium, in the absence of antibiotic, and incubated for a short time. Plasmid replication and expression can then get started, so that when the cells are plated out and encounter the antibiotic, they will already have synthesized sufficient resistance enzymes to be able to survive (Figure 5.5).

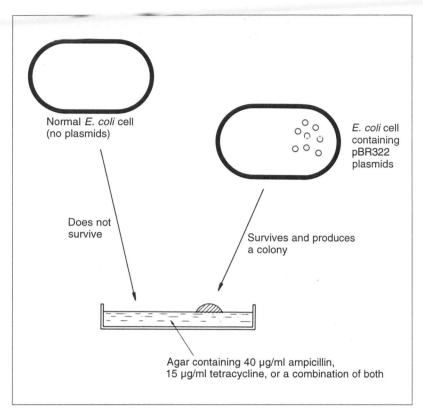

Figure 5.4 Selecting cells that contain pBR322 plasmids by plating on to agar medium containing ampicillin and/or tetracycline.

Normal *E. coli* cell
(no plasmids)

E. coli cell
containing
pBR322
plasmids

Does not
survive

Survives and produces
a colony

Agar containing 40 µg/ml ampicillin,
15 µg/ml tetracycline, or a combination of both

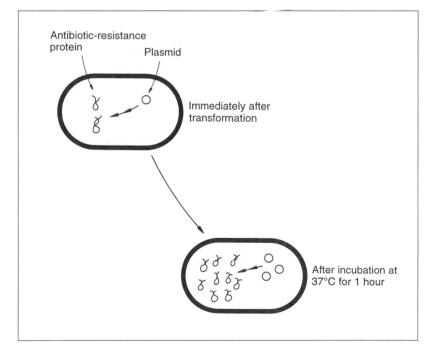

Figure 5.5 Phenotypic expression. Incubation at 37°C for 1 h before plating out improves the survival of the transformants on selective medium, because the bacteria have had time to begin synthesis of the antibiotic resistance enzymes.

Antibiotic-resistance
protein

Plasmid

Immediately after
transformation

After incubation at
37°C for 1 hour

Plating on to a selective medium enables transformants to be distinguished from non-transformants. The next problem is to determine which of the transformed colonies comprise cells that contain recombinant DNA molecules, and which contain self-ligated vector molecules (see Figure 5.2). With most cloning vectors insertion of a DNA fragment into the plasmid destroys the integrity of one of the genes present on the molecule. **Recombinants** can therefore be identified because the characteristic coded by the inactivated gene is no longer displayed by the host cells (Figure 5.6). The general principles of **insertional inactivation** are illustrated by a typical cloning experiment using pBR322 as the vector.

Figure 5.6 Insertional inactivation. (a) The normal, non-recombinant vector molecule carries a gene whose product confers a selectable or identifiable characteristic on the host cell. (b) This gene is disrupted when new DNA is inserted into the vector; as a result the recombinant host does not display the relevant characteristic.

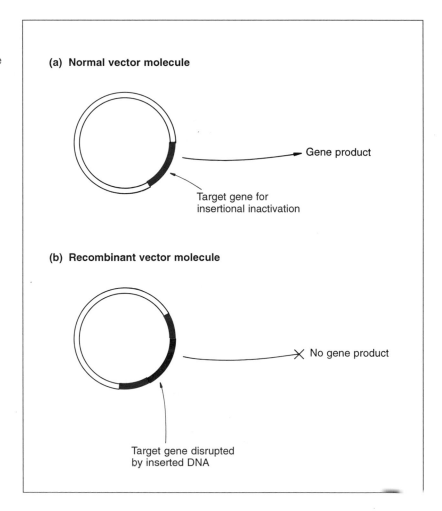

(a) **Normal vector molecule**

Gene product

Target gene for
insertional inactivation

(b) **Recombinant vector molecule**

No gene product

Target gene disrupted
by inserted DNA

5.2.1 Recombinant selection with pBR322 – insertional inactivation of an antibiotic resistance gene

pBR322 has several unique restriction sites that can be used to open up the vector before insertion of a new DNA fragment (Figure 5.7(a)). *Bam*HI, for example, cuts pBR322 at just one position, within the cluster of genes that code for resistance to tetracycline. A recombinant pBR322 molecule, one that carries an extra piece of DNA in the *Bam*HI site (Figure 5.7(b)), is no longer able to confer tetracycline resistance on its host, as one of the necessary genes is now disrupted by the inserted DNA. Cells containing this recombinant pBR322 molecule will still be resistant to ampicillin, but sensitive to tetracycline (amp^{R}tet^S).

Screening for pBR322 recombinants is performed in the following way. After transformation the cells are plated on to ampicillin medium and incubated until colonies appear (Figure 5.8(a)). All of these colonies are transformants (remember, untransformed cells are amp^S so do not produce colonies on the selective medium), but

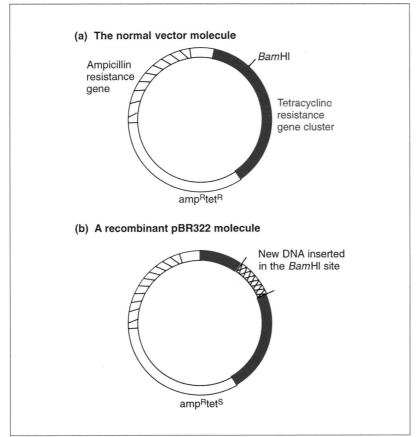

(a) **The normal vector molecule**

Ampicillin resistance gene

*Bam*HI

Tetracycline resistance gene cluster

amp^{R}tet^R

(b) **A recombinant pBR322 molecule**

New DNA inserted in the *Bam*HI site

amp^{R}tet^S

Figure 5.7 The cloning vector pBR322. (a) The normal vector molecule. (b) A recombinant molecule containing an extra piece of DNA inserted into the *Bam*HI site. For a more detailed map of pBR322 see Figure 6.1.

Figure 5.8 Screening for pBR322 recombinants by insertional inactivation of the tetracycline resistance gene. (a) Cells are plated on to ampicillin agar: all the transformants produce colonies. (b) The colonies are replica-plated on to tetracycline medium. (c) The colonies that grow on tetracycline medium are amp^{R}tet^R and therefore non-recombinants. Recombinants (amp^{R}tet^S) do not grow, but their position on the ampicillin plate is now known.

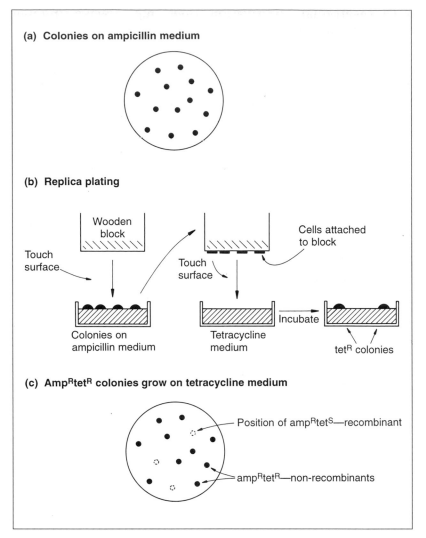

only a few contain recombinant pBR322 molecules, most contain the normal, self-ligated plasmid. To identify the recombinants the colonies are **replica-plated** on to agar medium that contains tetracycline (Figure 5.8(b)). After incubation, some of the original colonies will regrow, others will not (Figure 5.8(c)). Those that do grow consist of cells that carry the normal pBR322 with no inserted DNA and therefore a functional tetracycline resistance gene cluster (amp^{R}tet^R). The colonies that do not grow on tetracycline agar are recombinants (amp^{R}tet^S); once their positions are known samples for further study can be recovered from the original ampicillin agar plate.

5.2.2 Insertional inactivation does not always involve antibiotic resistance

Although insertional inactivation of an antibiotic resistance gene provides a convenient means of recombinant identification, several important plasmid cloning vectors make use of a different system. An example is pUC8 (Figure 5.9(a)), which carries the ampicillin resistance gene and a gene called *lacZ'*, which codes for part of the enzyme β-galactosidase. Cloning with pUC8 involves insertional inactivation of the *lacZ'* gene, with recombinants identified because of their inability to synthesize β-galactosidase (Figure 5.9(b)).

β-galactosidase is one of a series of enzymes involved in the breakdown of lactose to glucose plus galactose. It is normally coded by the gene *lacZ*, which resides on the *E. coli* chromosome. Some strains of *E. coli* have a modified *lacZ* gene, one that lacks the segment referred to as *lacZ'* and coding for the α-peptide portion of β-galactosidase (Figure 5.10(a)). These mutants can synthesize

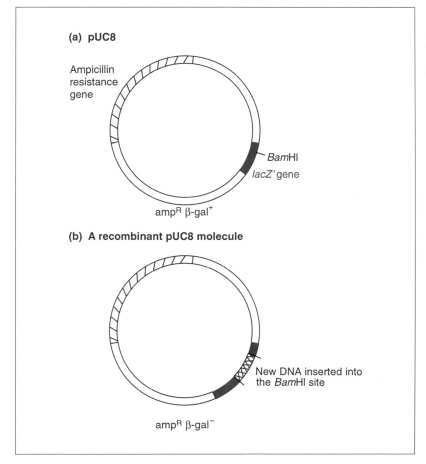

Figure 5.9 The cloning vector pUC8. (a) The normal vector molecule. (b) A recombinant molecule containing an extra piece of DNA inserted into the *Bam*HI site. For more detailed maps of pUC8 see Figures 6.3 and 6.4.

Figure 5.10 The rationale behind insertional inactivation of the *lacZ'* gene carried by pUC8. (a) The bacterial and plasmid genes complement each other to produce a functional β-galactosidase molecule. (b) Recombinants are screened by plating on to agar containing X-gal + IPTG.

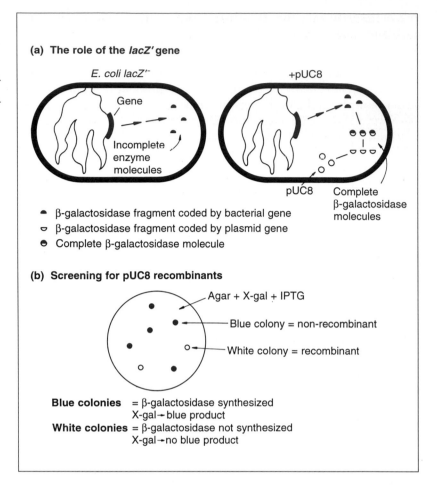

(a) The role of the *lacZ'* gene

E. coli lacZ'⁻   +pUC8

Gene

Incomplete enzyme molecules

pUC8   Complete β-galactosidase molecules

- β-galactosidase fragment coded by bacterial gene
- β-galactosidase fragment coded by plasmid gene
- Complete β-galactosidase molecule

(b) Screening for pUC8 recombinants

Agar + X-gal + IPTG

Blue colony = non-recombinant

White colony = recombinant

Blue colonies = β-galactosidase synthesized
X-gal → blue product
White colonies = β-galactosidase not synthesized
X-gal → no blue product

the enzyme only when they harbour a plasmid, such as pUC8, that carries the missing *lacZ'* segment of the gene.

A cloning experiment with pUC8 involves selection of transformants on ampicillin agar followed by screening for β-galactosidase activity to identify recombinants. Cells that harbour a normal pUC8 plasmid are amp^R and able to synthesize β-galactosidase (Figure 5.9(a)); recombinants are also amp^R but unable to make β-galactosidase (Figure 5.9(b)).

Screening for β-galactosidase presence or absence is in fact quite easy. Rather than assay for lactose being split to glucose and galactose, a slightly different reaction catalysed by the enzyme is tested for. This involves a lactose analogue called X-gal (5-bromo-4-chloro-3-indolyl-β-D-galactopyranoside) which is broken down by β-galactosidase to a product that is coloured deep blue. If X-gal (plus an inducer of the enzyme such as isopropyl-thiogalactoside, IPTG) is added to the agar, along with ampicillin, then non-recombinant colonies, the cells of which synthesize β-galactosidase, will be coloured blue, whereas recombinants with a disrupted *lacZ'* gene and unable to make β-galactosidase, will be white. This system, which is called **Lac selection**, is summarized in Figure 5.10(b)).

5.3 INTRODUCTION OF PHAGE DNA INTO BACTERIAL CELLS

There are two different methods by which a recombinant DNA molecule constructed with a phage vector can be introduced into a bacterial cell: transfection and *in vitro* **packaging**.

(a) Transfection   This process is equivalent to transformation, the only difference being that phage DNA rather than a plasmid is involved. Just as with a plasmid, the purified phage DNA, or recombinant phage molecule, is mixed with competent *E. coli* cells and DNA uptake induced by heat-shock.

(b) *In vitro* packaging   Transfection with λ DNA molecules is not a very efficient process when compared with the infection of a culture of cells with mature λ phage particles. It would therefore be useful if recombinant λ molecules could be packaged into the λ head-and-tail structures in the test-tube.

This may sound difficult but in fact is relatively easy to achieve. Packaging requires a number of different proteins coded by the λ genome, but these can be prepared at a high concentration from cells infected with defective λ phage strains. Two different systems are in use. With the single-strain system the defective λ phage carries a mutation in the *cos* sites, so that these are not recognized by the endonuclease that normally cleaves the λ concatamers during phage replication (p. 23). The defective phage cannot therefore replicate, though it does direct synthesis of all the proteins needed for packaging. The proteins accumulate in the bacterium and can be purified from cultures of *E. coli* infected with the mutated λ, and used for *in vitro* packaging of recombinant λ molecules (Figure 5.11(a)).

With the second system two defective λ strains are needed. Both of these strains carry a mutation in a gene for one of the components of the phage protein coat: with one strain the mutation is in gene *D*, and with the second strain it is in gene *E* (see Figure 2.9). Neither strain is able to complete an infection cycle in *E. coli* as in the absence of the product of the mutated gene the complete capsid structure cannot be made. Instead the products of all the other coat protein genes accumulate (Figure 5.11(b)). An *in vitro* packaging mix can therefore be prepared by combining lysates of two

Figure 5.11 *In vitro* packaging.
(a) Synthesis of λ capsid proteins by
E. coli strain SMR10, which carries a
λ phage that has defective *cos* sites.
(b) Synthesis of incomplete sets of λ
capsid proteins by *E. coli* strains
BHB2688 and BHB2690. (c) A mixture of cell lysates provides the complete set of capsid proteins and can
package λ DNA molecules in the
test-tube.

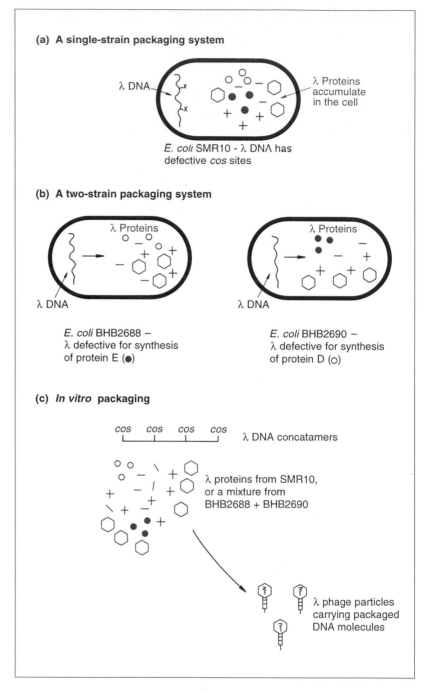

cultures of cells, one infected with the λ *D⁻* strain, the other
infected with the *E⁻* strain. The mixture now contains all the necessary components and will package recombinant λ molecules into
mature phage particles (Figure 5.11(c)). With both systems, pack-

aged molecules can be introduced into *E. coli* cells simply by adding the assembled phage to the bacterial culture and allowing the normal λ infective process to take place.

5.3.1 Phage infection is visualized as plaques on an agar medium

The final-stage of the phage infection cycle is cell lysis (p. 19). If infected cells are spread on to a solid agar medium immediately after addition of the phage, or immediately after transfection with phage DNA, then cell lysis can be visualized as **plaques** on a lawn of bacteria (Figure 5.12(a)). Each plaque is a zone of clearing produced as the phage lyse the cells and move on to infect and eventually lyse the neighbouring bacteria (Figure 5.12(b)).

Both λ and M13 form plaques. With λ these are true plaques, produced by cell lysis. However, M13 plaques are slightly

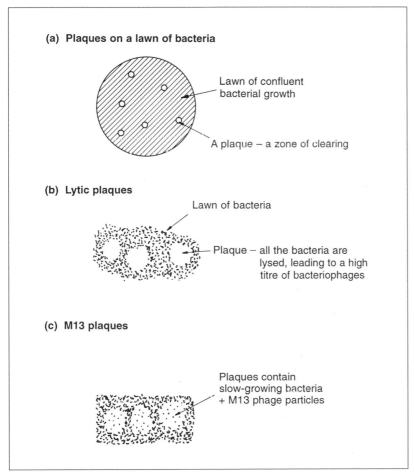

(a) **Plaques on a lawn of bacteria**

Lawn of confluent bacterial growth

A plaque – a zone of clearing

(b) **Lytic plaques**

Lawn of bacteria

Plaque – all the bacteria are lysed, leading to a high titre of bacteriophages

(c) **M13 plaques**

Plaques contain slow-growing bacteria + M13 phage particles

Figure 5.12 Bacteriophage plaques. (a) The appearance of plaques on a lawn of bacteria. (b) Plaques produced by a phage that lyses the host cell (e.g. λ in the lytic infection cycle); the plaques contain lysed cells plus many phage particles. (c) Plaques produced by M13; these plaques contain slow-growing bacteria plus many M13 phage particles.

different as M13 does not lyse the host cells (p. 20). Instead M13 causes a decrease in the growth rate of infected cells sufficient to produce a zone of relative clearing on a bacterial lawn. Although not a true plaque, these zones of clearing are visually identical to normal phage plaques (Figure 5.12(c)).

The end result of a gene cloning experiment using a λ or M13 vector is therefore an agar plate covered in phage plaques. Each plaque is derived from a single transfected or infected cell and therefore contains identical phage particles. These may contain self-ligated vector molecules, or they may be recombinants.

5.4 IDENTIFICATION OF RECOMBINANT PHAGES

A variety of ways of distinguishing recombinant plaques have been devised, the following being the most important.

5.4.1 Insertional inactivation of a *lacZ'* gene carried by the phage vector

All M13 cloning vectors (p. 115), as well as several λ vectors, carry a copy of the *lacZ'* gene. Insertion of new DNA into this gene inactivates β-galactosidase synthesis, just as with the plasmid vector pUC8. Recombinants are distinguished by plating cells on to X-gal agar: plaques comprising normal phage are blue, recombinant plaques are clear (Figure 5.13(a)).

5.4.2 Insertional inactivation of the λ *c*I gene

Several types of λ cloning vector have unique restriction sites in the *c*I gene (map position 38 on Figure 2.9). Insertional inactivation of this gene causes a change in plaque morphology. Normal plaques appear 'turbid', whereas recombinants with a disrupted *c*I gene are 'clear' (Figure 5.13(b)). The difference is readily apparent to the experienced eye.

5.4.3 Selection using the Spi phenotype

λ phage cannot normally infect *E. coli* cells that already possess an integrated form of a related phage called P2. λ is therefore said to be Spi+ (sensitive to P2 prophage inhibition). Some λ cloning vectors are designed so that insertion of new DNA causes a change from Spi+ to Spi−, thus the recombinants can infect cells that carry P2 prophages. Such cells are used as the host for cloning experiments with these vectors; only recombinants are Spi− so only recombinants form plaques (Figure 5.13(c)).

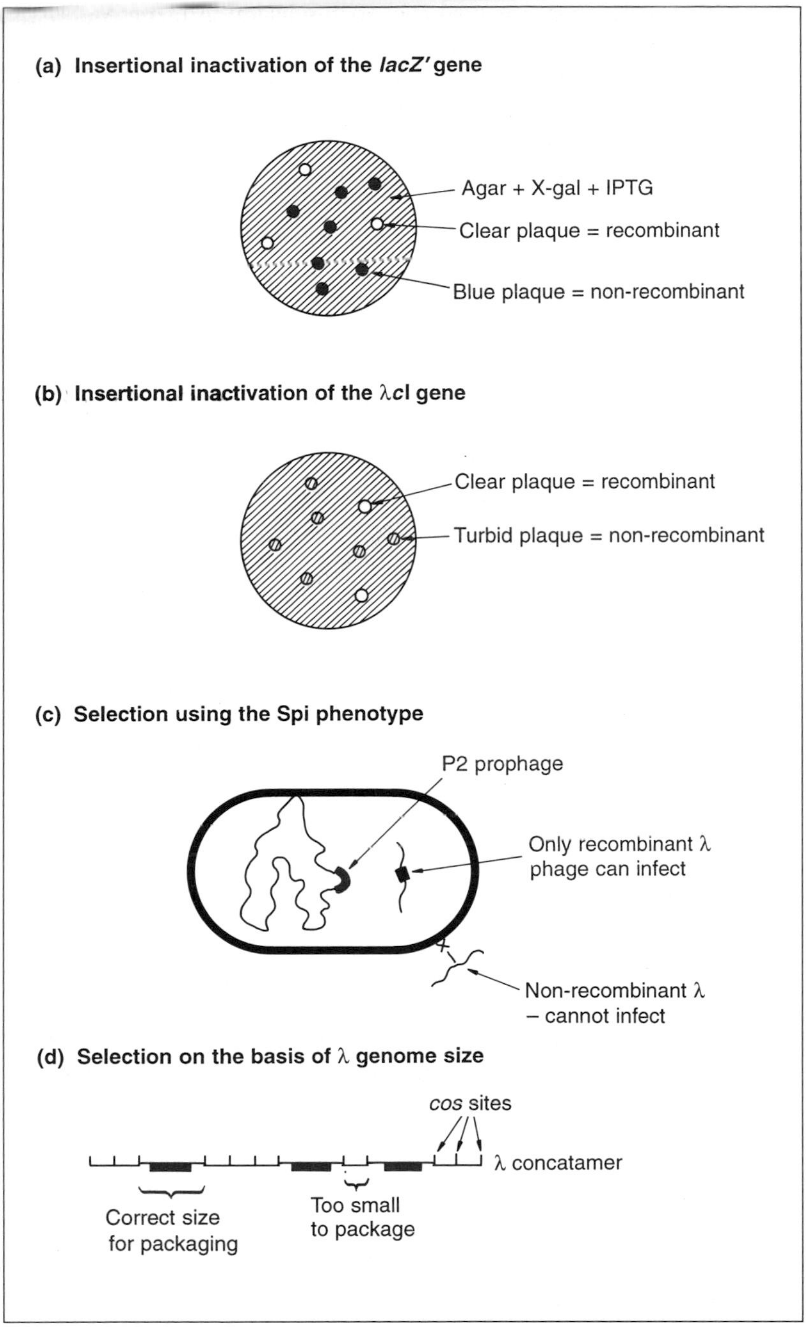

Figure 5.13 Strategies for the selection of recombinant phage.

(a) Insertional inactivation of the *lacZ'* gene

Agar + X-gal + IPTG

Clear plaque = recombinant

Blue plaque = non-recombinant

(b) Insertional inactivation of the λcI gene

Clear plaque = recombinant

Turbid plaque = non-recombinant

(c) Selection using the Spi phenotype

P2 prophage

Only recombinant λ phage can infect

Non-recombinant λ – cannot infect

(d) Selection on the basis of λ genome size

cos sites

λ concatamer

Correct size for packaging

Too small to package

5.4.4 Selection on the basis of λ genome size

The λ packaging system, which assembles the mature phage particles, can only insert DNA molecules of between 37 and 52 kb into the head structure. Anything less than 37 kb will not be packaged.

Many λ vectors have been constructed by deleting large segments of the λ DNA molecule (p. 123) and so are less than 37 kb in length. These will only package into mature phage particles after extra DNA has been inserted, bringing the total genome size up to 37 kb or more (Figure 5.13(d)). Therefore with these vectors only recombinant phage are able to replicate.

5.5 TRANSFORMATION OF NON-BACTERIAL CELLS

Ways of introducing DNA into yeast, fungi, animals and plants are also needed if these organisms are to be used as the hosts for gene cloning. In general terms, soaking cells in salt is effective only with a few species of bacteria, although treatment with lithium chloride or lithium acetate does enhance DNA uptake by yeast cells, and is frequently used in the transformation of *Saccharomyces cerevisiae*. However, for most higher organisms more sophisticated methods are needed.

5.5.1 Transformation of individual cells

With most organisms the main barrier to DNA uptake is the cell wall. Cultured animal cells, which usually lack cell walls, are easily transformed, especially if the DNA is precipitated on to the cell surface with calcium phosphate (Figure 5.14(a)). For other types of cell the answer is often to remove the cell wall. Enzymes that degrade yeast, fungal and plant cell walls are available, and under the right conditions intact **protoplasts** can be obtained (Figure 5.14(b)). Protoplasts generally take up DNA quite readily; alternatively transformation can be stimulated by special techniques such as **electroporation**, during which the cells are subjected to a short electrical pulse, thought to induce the transient formation of pores in the cell membrane through which the DNA molecules can enter the cell. After transformation the protoplasts are washed to remove the degradative enzymes and the cell wall spontaneously reforms.

In contrast to the transformation systems described so far there are two physical methods for introducing DNA into cells. The first of these is **microinjection**, which makes use of a very fine pipette to inject DNA molecules directly into the nucleus of the cells to be transformed (Figure 5.15(a)). This technique was initially applied to animal cells but has subsequently been successful with plant cells. The second method involves bombardment of the cells with high-velocity microprojectiles, usually particles of gold or tungsten that have been coated with DNA. These microprojectiles are fired at the cells from a particle gun (Figure 5.15(b)). This unusual

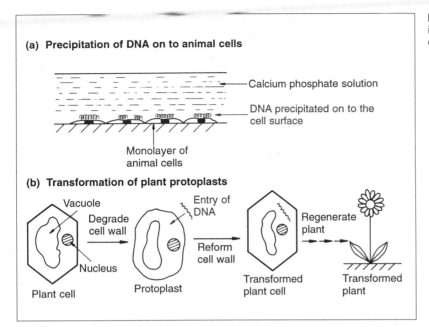

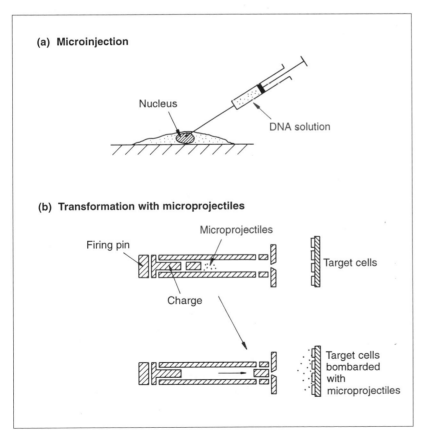

Figure 5.14 Strategies for introducing new DNA into animal and plant cells.

Figure 5.15 Two physical methods for introducing DNA into cells.

technique is termed **biolistics** and has been used with a number of different types of cell.

5.5.2 Transformation of whole organisms

With animals and plants the desired end-product may not be transformed cells, but a transformed organism. Plants are relatively easy to regenerate from cultured cells, though problems have been experienced in developing regeneration procedures for monocotyledonous species such as cereals and grasses. A single transformed plant cell can therefore give rise to a transformed plant, which will carry the cloned DNA in every cell, and will pass the cloned DNA on to its progeny following flowering and seed formation (Figure 7.13(b)). Animals of course cannot be regenerated from cultured cells, so obtaining transformed animals requires a rather more subtle approach. The standard technique with mammals such as mice is to remove fertilized eggs from the oviduct, to microinject DNA, and then to re-implant the transformed cells into the mother's reproductive tract.

FURTHER READING

Cohen, S. N. *et al.* (1972) Nonchromosomal antibiotic resistance in bacteria: genetic transformation of *Escherichia coli* by R-factor DNA. *Proceedings of the National Academy of Sciences USA*, **69**, 2110–14 – transformation of a bacterium with a plasmid.

Mandel, M. and Higa, A. (1970) Calcium-dependent bacteriophage DNA infection. *Journal of Molecular Biology*, **53**, 154–62 – transfection.

Hohn, B. and Murray, K. (1977) Packaging recombinant DNA molecules into bacteriophage particles *in vitro*. *Proceedings of the National Academy of Sciences USA*, **74**, 3259–63 – *in vitro* packaging.

Calvin, N. M. and Hanawalt, P. C. (1988) High-efficiency transformation of bacterial cells by electroporation. *Journal of Bacteriology*, **170**, 2796–801.

Capecchi, M. R. (1980) High efficiency transformation by direct microinjection of DNA into cultured mammalian cells. *Cell*, **22**, 479–88.

Klein, T. M. *et al.* (1987) High velocity microprojectiles for delivering nucleic acids into living cells. *Nature*, **327**, 70–3 – biolistics.

Hammer, R. E., Pursel, V. G., Rexroad, C. E., Wall, R. J., Bolt, D. J., Ebert, K. M., Palmiter, R. D. and Brinster, R. L. (1985) Production of transgenic rabbits, sheep and pigs by microinjection. *Nature*, **315**, 680–3

Cloning vectors for *E. coli*

6

The basic experimental techniques involved in gene cloning have now been described. In Chapters 3, 4 and 5 we have seen how DNA can be purified from cell extracts, how recombinant DNA molecules can be constructed in the test-tube, how DNA molecules can be reintroduced into living cells, and how recombinant clones can be distinguished. Now we must look more closely at the cloning vector itself, in order to consider the range of vectors available to the molecular biologist, and to understand the properties and uses of each individual type.

The greatest variety of cloning vectors exist for use with *E. coli* as the host organism. This is not surprising in view of the central role that this bacterium has played in basic research over the last 50 years. The tremendous wealth of information that exists concerning the microbiology, biochemistry and genetics of *E. coli* has meant that virtually all fundamental studies of gene structure and function have been carried out with this bacterium as the experimental organism. In recent years, gene cloning and molecular biological research have become mutually synergistic – breakthroughs in gene cloning have acted as a stimulus to research, and the needs of research have spurred on the development of new, more sophisticated cloning vectors.

In this chapter the most important types of *E. coli* cloning vector will be described, and the specific uses of representative molecules outlined. In Chapter 7, cloning vectors for yeast, fungi, plants and animals will be considered.

6.1 CLONING VECTORS BASED ON *E. COLI* PLASMIDS

The simplest cloning vectors, and the ones in most widespread use in gene cloning, are those based on small bacterial plasmids. A large number of different plasmid vectors are available for use with *E. coli*, many obtainable from commercial suppliers. They combine ease of purification with desirable properties such as high transformation efficiency, convenient selectable markers for transformants and recombinants, and the ability to clone reasonably large (up to about 8 kb) pieces of DNA. Most 'routine' gene cloning experiments make use of one or other of these plasmid vectors.

One of the first vectors to be developed, and still one of the most popular today, is pBR322, which was introduced in Chapter 5 to illustrate the general principles of transformant selection and recombinant identification (p. 95). We will begin our study of *E. coli* plasmid vectors by looking more closely at pBR322.

6.1.1 The nomenclature of plasmid cloning vectors

The name 'pBR322' conforms with the standard rules for vector nomenclature.

'p' indicates that this is indeed a plasmid
'BR' identifies the laboratory in which the vector was originally constructed (BR stands for Bolivar and Rodriguez, the two researchers who developed pBR322)
'322' distinguishes this plasmid from others developed in the same laboratory (there are also plasmids called pBR325, pBR327, pBR328, etc.)

6.1.2 The useful properties of pBR322

The genetic and physical map of pBR322 (Figure 6.1) gives an indication of why this plasmid has been such a popular cloning vector.

The first useful feature of pBR322 is its size. In Chapter 2 it was stated that a cloning vector ought to be less than 10 kb in size, to avoid problems such as DNA breakdown during purification.

Figure 6.1 A map of pBR322 showing the positions of the ampicillin resistance (amp^R) and tetracycline resistance (tet^R) genes, the origin of replication (ori) and a selection of the most important restriction sites.

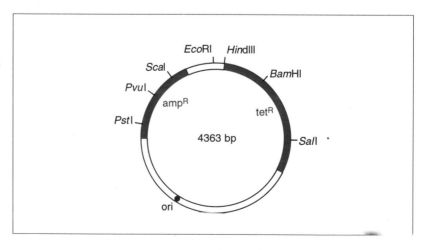

pBR322 is 4363 bp, which means that not only can the vector itself be purified with ease, but so can recombinant DNA molecules constructed with it. Even with 6 kb of additional DNA, a recombinant pBR322 molecule will still be a manageable size.

The second feature of pBR322 is that, as described in Chapter 5, it carries two sets of antibiotic resistance genes. Either ampicillin or tetracycline resistance can be used as a selectable marker for cells containing the plasmid, and each marker gene includes unique restriction sites that can be used in cloning experiments. Insertion of new DNA into pBR322 that has been restricted with *Pst*I, *Pvu*I or *Sca*I inactivates the amp^R gene, and insertion using any one of eight restriction endonucleases (notably *Bam*HI and *Hin*dIII) inactivates tetracycline resistance. This great variety of restriction sites that can be used for insertional inactivation means that pBR322 can be used to clone DNA fragments with any of several kinds of sticky end.

A third advantage of pBR322 is that it has a reasonably high copy number. Generally there are about 15 molecules present in a transformed *E. coli* cell, but this number can be increased, up to 1000–3000, by plasmid amplification in the presence of a protein synthesis inhibitor such as chloramphenicol (p. 43). An *E. coli* culture will therefore provide a good yield of recombinant pBR322 molecules.

6.1.3 The pedigree of pBR322

The remarkable convenience of pBR322 as a cloning vector did not arise by chance. The plasmid was in fact designed in such a way that the final construct would possess these desirable properties. An outline of the scheme used to construct pBR322 is shown in Figure 6.2(a). It can be seen that its production was a tortuous business that required full and skilful use of the DNA manipulative techniques described in Chapter 4. A summary of the result of these manipulations is provided in Figure 6.2(b), from which it can be seen that pBR322 in fact comprises DNA derived from three different naturally occurring plasmids. The amp^R gene originally resided on the plasmid R1, a typical antibiotic resistance plasmid that occurs in natural populations of *E. coli* (p. 17). The tet^R gene is derived from R6-5, a second antibiotic resistance plasmid, and the replication origin of pBR322, which directs multiplication of the vector in host cells, is originally from pMB1, which is closely related to the colicin-producing plasmid ColE1 (p. 17).

6.1.4 Other typical *E. coli* plasmid cloning vectors

pBR322 was developed in the late 1970s, the first research paper describing its use being published in 1977. Since then many other

Figure 6.2 The pedigree of pBR322. (a) The manipulations involved in construction of pBR322. (b) A summary of the origins of pBR322.

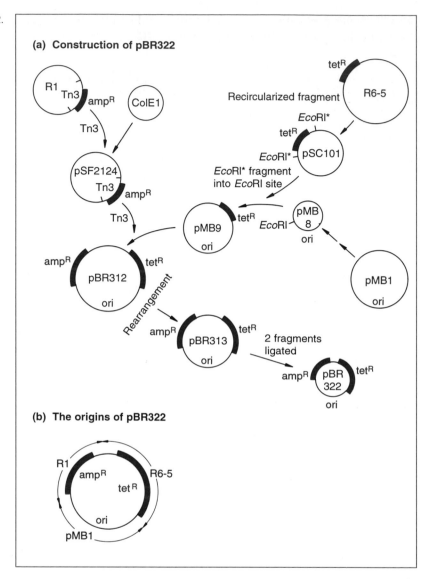

plasmid cloning vectors have been constructed, the majority of these derived from pBR322 by manipulations similar to those summarized in Figure 6.2(a). It would be pointless to attempt to describe all of these vectors, especially as many are variations on a common theme. Three additional examples will suffice to illustrate the most important features displayed by the range of plasmid cloning vectors available to today's genetic engineers.

(a) pBR327 – a higher copy number plasmid (Figure 6.3(a))
pBR327 was constructed by removing a 1089 bp segment from pBR322. This deletion left the amp^R and tet^R genes intact, but

changed the replicative and conjugative abilities of the resulting plasmid. As a result pBR327 differs from pBR322 in two important ways.

1. pBR327 has a higher copy number than pBR322, being present at about 30–45 molecules per *E. coli* cell. This is not of great relevance as far as plasmid yield is concerned, as both plasmids can be amplified to copy numbers greater than 1000. However, the higher copy number of pBR327 in normal cells makes this vector more suitable if the aim of the experiment is to study the function of the cloned gene. In these cases gene dosage becomes important, as the more copies there are of a cloned gene, the more likely it is that the effect of the cloned gene on the host cell will be detectable. pBR327, with its high copy number, is therefore a better choice than pBR322 for this kind of work.
2. The deletion also destroys the conjugative ability of pBR322, making pBR327 a non-conjugative plasmid that cannot direct its own transfer to other *E. coli* cells. This is important for **biological containment**, averting the possibility of a recombinant pBR327 molecule escaping from the test-tube and colonizing bacteria in the gut of a careless molecular biologist. In contrast, pBR322 could theoretically be passed to natural populations of *E. coli* by conjugation, though in fact pBR322 also has safeguards

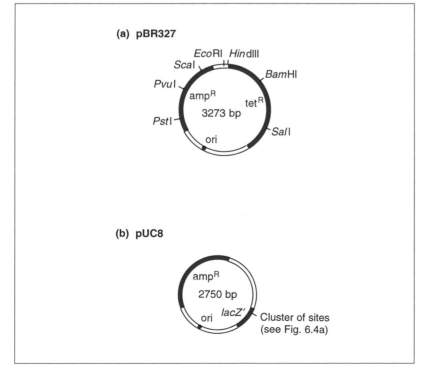

(a) pBR327

(b) pUC8

Figure 6.3 Two *E. coli* plasmid cloning vectors.

(though less sophisticated ones) to minimize the chances of this happening. pBR327 is, however, preferable if the cloned gene is potentially harmful should an accident occur.

(b) pUC8 – a Lac selection plasmid (Figure 6.3(b)) This vector was mentioned in Chapter 5 when identification of recombinants by insertional inactivation of the β-galactosidase gene was described (p. 97). pUC8 is derived from pBR322, although only the replication origin and the amp^R gene remain. The nucleotide sequence of the amp^R gene has been changed so that it no longer contains the unique restriction sites; all these cloning sites are now clustered into a short segment of the *lacZ'* gene carried by pUC8.

pUC8 has three important advantages that have led to it becoming one of the most popular *E. coli* cloning vectors. The first of these is fortuitous: the manipulations involved in construction of pUC8 were accompanied by a chance mutation, within the origin of replication, that results in the plasmid having a copy number of 500–700 even before amplification. This has a significant effect on the yield of cloned DNA obtainable from *E. coli* cells transformed with recombinant pUC8 plasmids. The second advantage is that identification of recombinant cells can be achieved by a single-step process, by plating on to agar medium containing ampicillin plus X-gal (p. 98). With both pBR322 and pBR327, selection of recombinants is a two-step procedure, requiring replica-plating from one antibiotic medium to another (p. 96). A cloning experiment with pUC8 can therefore be carried out in half the time needed with pBR322 or pBR327.

The third advantage of pUC8 lies with the clustering of the restriction sites, which allows a DNA fragment with two different sticky ends (say *Eco*RI at one end and *Bam*HI at the other) to be cloned without resorting to additional manipulations such as linker attachment (Figure 6.4(a)). Other pUC vectors carry different combinations of restriction sites and provide even greater flexibility in the types of DNA fragment that can be cloned (Figure 6.4(b)). Furthermore, the restriction site clusters in these vectors are the same as the clusters in the equivalent M13mp series of vectors (p. 119). DNA cloned into a member of the pUC series can therefore be transferred directly to its M13mp counterpart, in which form it can be analysed by DNA sequencing or *in vitro* mutagenesis (Figure 6.4(c); see p. 193 and 223 for descriptions of these procedures).

(c) pGEM3Z – *in vitro* transcription of cloned DNA (Figure 6.5(a)) pGEM3Z is very similar to a pUC vector: it carries the amp^R and *lacZ'* genes, the latter containing a cluster of restriction sites, and it is almost exactly the same size. The distinction is that pGEM3Z has two additional, short pieces of DNA, each of which

(a) Restriction sites in pUC8

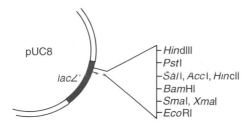

pUC8

lacZ'

- *Hind*III
- *Pst*I
- *Sal*I, *Acc*I, *Hinc*II
- *Bam*HI
- *Sma*I, *Xma*I
- *Eco*RI

(b) Restriction sites in pUC18

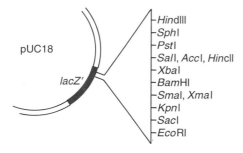

pUC18

lacZ'

- *Hind*III
- *Sph*I
- *Pst*I
- *Sal*I, *Acc*I, *Hinc*II
- *Xba*I
- *Bam*HI
- *Sma*I, *Xma*I
- *Kpn*I
- *Sac*I
- *Eco*RI

(c) Shuttling a DNA fragment from pUC8 to M13mp8

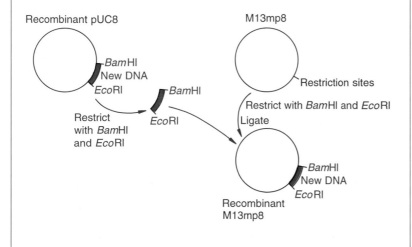

Recombinant pUC8

*Bam*HI
New DNA
*Eco*RI

Restrict with *Bam*HI and *Eco*RI

*Bam*HI

*Eco*RI

M13mp8

Restriction sites

Restrict with *Bam*HI and *Eco*RI

Ligate

*Bam*HI
New DNA
*Eco*RI

Recombinant M13mp8

Figure 6.4 The pUC plasmids. (a) The restriction site cluster in the *lacZ'* gene of pUC8. (b) The restriction site cluster in pUC18. (c) Shuttling a DNA fragment from pUC8 to M13mp8.

Figure 6.5 pGEM3Z. (a) Map of the vector. (b) *In vitro* RNA synthesis. R, cluster of restriction sites for *Eco*RI, *Sac*I, *Kpn*I, *Ava*I, *Sma*I, *Bam*HI, *Xba*I, *Sal*I, *Acc*I, *Hinc*II, *Pst*I, *Sph*I and *Hind*III.

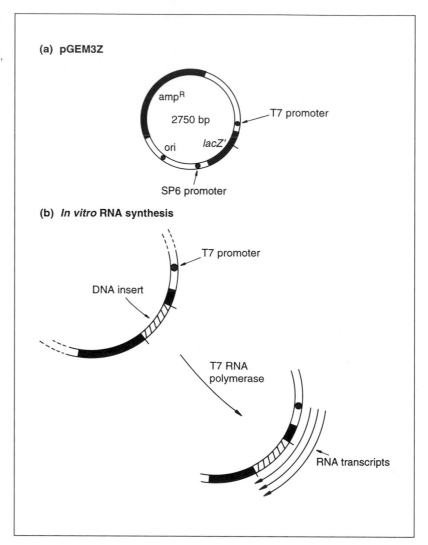

(a) pGEM3Z

amp^R

2750 bp

ori   *lacZ'*

T7 promoter

SP6 promoter

(b) *In vitro* RNA synthesis

T7 promoter

DNA insert

T7 RNA polymerase

RNA transcripts

acts as the recognition site for attachment of an RNA polymerase enzyme. These two **promoter** sequences lie on either side of the cluster of restriction sites used for introduction of new DNA into the pGEM3Z molecule. This means that if a recombinant pGEM37 molecule is mixed with purified RNA polymerase in the test-tube, transcription occurs and RNA copies of the cloned fragment are synthesized (Figure 6.5(b)). The RNA that is produced could be used as a hybridization probe (p. 168), or might be required for experiments aimed at studying RNA processing (e.g. the removal of introns) or protein synthesis.

The promoters carried by pGEM3Z and other vectors of this type are not the standard sequences recognized by the *E. coli* RNA polymerase. Instead, one of the promoters is specific for the RNA polymerase coded by T7 bacteriophage and the other for the RNA polymerase of SP6 phage. These RNA polymerases are synthesized during infection of *E. coli* with one or other of the phages and are responsible for transcribing the phage genes. They are chosen for use in *in vitro* transcription as they are very active enzymes (remember that the entire lytic infection cycle takes only 20 minutes – p. 19) and are able to synthesize 1–2 µg of RNA per minute, substantially more than can be produced by the standard *E. coli* enzyme.

6.2 CLONING VECTORS BASED ON M13 BACTERIOPHAGE

The most essential requirement for any cloning vector is that it has a means of replicating in the host cell. For plasmid vectors this requirement is easy to satisfy, as relatively short DNA sequences are able to act as plasmid origins of replication, and most, if not all, of the enzymes needed for replication are provided by the host cell. Elaborate manipulations, such as those that resulted in pBR322 (Figure 6.2(a)), are therefore possible so long as the final construction has an intact, functional replication origin.

With bacteriophages such as M13 the situation as regards replication is more complex. Phage DNA molecules generally carry several genes that are essential for replication, including genes coding for components of the phage protein coat and phage-specific DNA replicative enzymes. Alteration or deletion of any of these genes will impair or destroy the replicative ability of the resulting molecule. There is therefore much less freedom to modify phage DNA molecules, and generally phage cloning vectors are only slightly different from the parent molecule.

The problems in constructing a phage cloning vector are illustrated by considering M13. The normal M13 genome is 6.4 kb in length, but most of this is taken up by ten closely packed genes (Figure 6.6), each essential for the replication of the phage. There is only a single, 507 nucleotide intergenic sequence into which new DNA could be inserted without disrupting one of these genes, and in fact this region includes the replication origin which must itself remain intact. Clearly there is only limited scope for modifying the M13 genome.

Nevertheless, it will be remembered that the great attraction of M13 is the opportunity it offers of obtaining single-stranded versions of cloned DNA (p. 26). This feature has acted as a stimulus for the development of M13 cloning vectors.

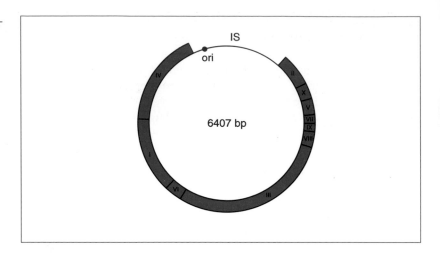

Figure 6.6 The M13 genome showing the positions of genes I to X.

6.2.1 Development of the cloning vector M13mp2

The first step in construction of an M13 cloning vector was to introduce the *lacZ'* gene into the intergenic sequence. This gave rise to M13mp1, which forms blue plaques on X-gal agar (Figure 6.7(a)).

M13mp1 does not possess any unique restriction sites in the *lacZ'* gene. It does, however, contain the hexanucleotide GGATTC, near the start of the gene. A single nucleotide change would make this GAATTC, which is an *Eco*RI site. This alteration was carried out using *in vitro* mutagenesis (p. 221), resulting in M13mp2 (Figure 6.7(b)). M13mp2 has a slightly altered *lacZ'* gene (the sixth codon now specifies asparagine instead of aspartic acid), but the β-galactosidase enzyme produced by cells infected with M13mp2 is still perfectly functional.

M13mp2 is the simplest M13 cloning vector. DNA fragments with *Eco*RI sticky ends can be inserted into the cloning site, and recombinants are distinguished as clear plaques on X-gal agar.

6.2.2 M13mp7 – symmetrical cloning sites

The next step in the development of M13 vectors was to introduce additional restriction sites into the *lacZ'* gene. This was achieved by synthesizing in the test-tube a short oligonucleotide, called a **polylinker**, that consists of a series of restriction sites and has *Eco*RI sticky ends (Figure 6.8(a)). This polylinker was inserted into the *Eco*RI site of M13mp2, to give M13mp7 (Figure 6.8(b)), a more complex vector with four possible cloning sites (*Eco*RI, *Bam*HI, *Sal*I and *Pst*I). The polylinker is designed so that it does not totally disrupt the *lacZ'* gene; a reading frame is maintained throughout the polylinker, and a functional, though altered, β-galactosidase enzyme is still produced.

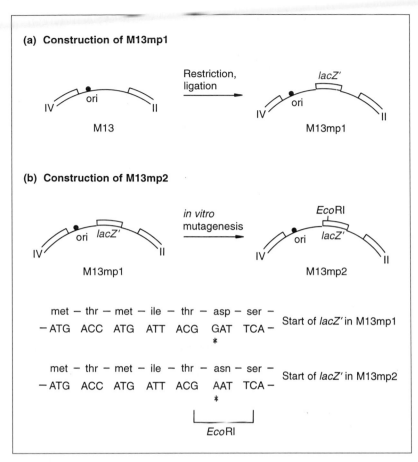

Figure 6.7 Construction of (a) M13mp1 and (b) M13mp2 from the wild-type M13 genome.

Figure 6.8 Construction of M13mp7. (a) The polylinker and (b) its insertion into the *Eco*RI site of M13mp2. Note that the *Sal*I restriction sites are also recognized by *Acc*I and *Hinc*II.

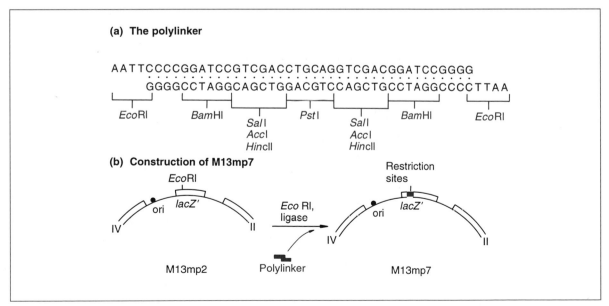

When M13mp7 is digested with either *Eco*RI, *Bam*HI or *Sal*I, a part or all of the polylinker is excised (Figure 6.9(a)). On ligation, in the presence of new DNA, one of three events may occur (Figure 6.9(b)).

1. New DNA is inserted.
2. The polylinker is reinserted.
3. The vector self-ligates without insertion.

Insertion of new DNA almost invariably prevents β-galactosidase production, so recombinant plaques are clear on X-gal agar (Figure 6.9(c)). Alternatively, if the polylinker is reinserted, and the original M13mp7 reformed, then blue plaques result. But what if the vector self-ligates, with neither new DNA nor the polylinker inserted? Once again the design of the polylinker comes into play. Whichever restriction site is used, self-ligation results in a functional *lacZ'* gene (Figure 6.9(c)), giving blue plaques. Selection is therefore unequivocal: only recombinant M13mp7 phage give rise to clear plaques.

A big advantage of M13mp7, with its symmetrical cloning sites, is that DNA inserted into either the *Bam*HI, *Sal*I or *Pst*I sites can be excised from the recombinant molecule using *Eco*RI (Figure

Figure 6.9 Cloning with M13mp7. See the text for details.

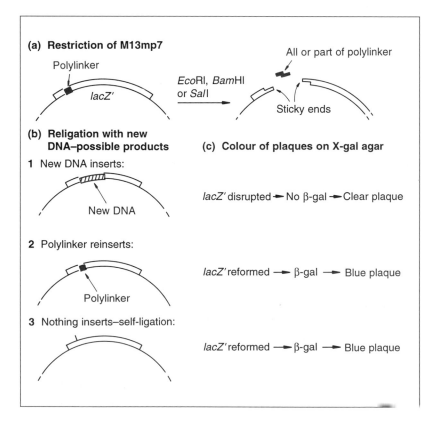

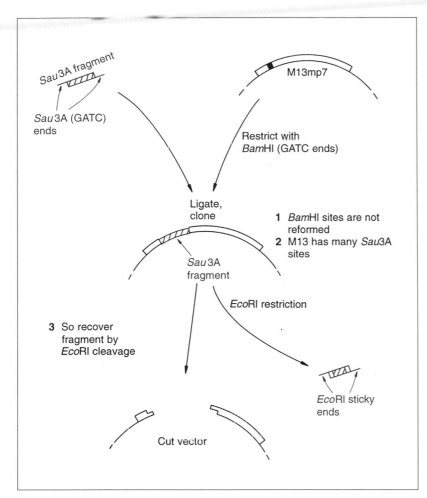

Figure 6.10 Recovery of cloned DNA from a recombinant M13mp7 molecule by restriction at the outer sites of the polylinker.

6.10). Very few vectors allow cloned DNA to be recovered so easily.

6.2.3 More complex M13 vectors

The latest M13 vectors have more complex polylinkers inserted into the *lacZ'* gene. An example is M13mp8 (Figure 6.11(a)), which is the counterpart of the plasmid pUC8 (p. 112). As with the plasmid vector, one advantage of M13mp8 is its ability to take DNA fragments with two different sticky ends.

A second feature is provided by the sister vector M13mp9 (Figure 6.11(b)), which has the same polylinker but in the reverse orientation. A DNA fragment cloned into M13mp8, if excised by double restriction, and then inserted into M13mp9, will now itself be in the reverse orientation (Figure 6.11(c)). This is important in DNA sequencing (p. 192), in which the nucleotide sequence is read

Figure 6.11 M13mp8 and M13mp9.

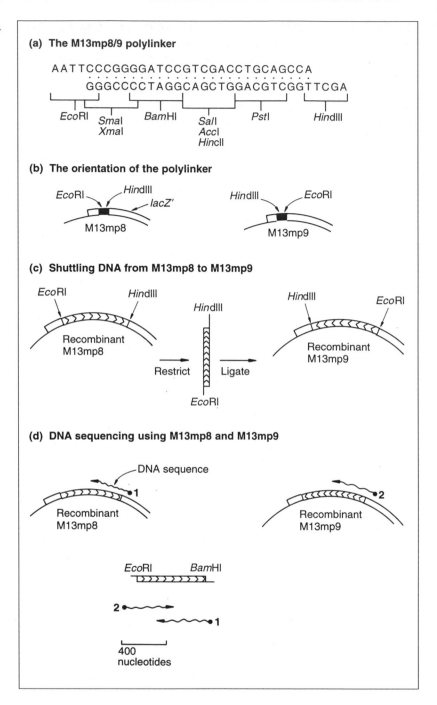

(a) **The M13mp8/9 polylinker**

```
AATTCCCGGGGATCCGTCGACCTGCAGCCA
· · · · · · · · · · · · · · · · · · · · · · · · · · · · · ·
    GGGCCCCTAGGCAGCTGGACGTCGGTTCGA
```

*EcoR*I
*Sma*I
*Xma*I
*Bam*HI
*Sal*I
*Acc*I
*Hinc*II
*Pst*I
*Hind*III

(b) **The orientation of the polylinker**

*EcoR*I   *Hind*III   — *lacZ'*
M13mp8

*Hind*III   *EcoR*I
M13mp9

(c) **Shuttling DNA from M13mp8 to M13mp9**

*EcoR*I   *Hind*III
Recombinant M13mp8

*Hind*III   Restrict   Ligate   *EcoR*I

*Hind*III   *EcoR*I
Recombinant M13mp9

(d) **DNA sequencing using M13mp8 and M13mp9**

— DNA sequence
●1
Recombinant M13mp8

●2
Recombinant M13mp9

*EcoR*I   *Bam*HI

2 ●

● 1

400 nucleotides

from one end of the polylinker into the inserted DNA fragment (Figure 6.11(d)). Only about 400 nucleotides can be read from one sequencing experiment; if the inserted DNA is longer than this, one end of the fragment will not be sequenced. The answer is to turn the fragment around, by excising and reinserting into the sister vector. A DNA sequencing experiment with this new clone will allow the nucleotide sequence at the other end of the fragment to be determined.

Other M13 vector pairs are also available. M13mp10/11 and M13mp18/19 are similar to M13mp8/9, but have different polylinkers and therefore different restriction sites.

6.2.4 Hybrid plasmid–M13 vectors

Although M13 vectors are very useful for the production of single-stranded versions of cloned genes they do suffer from one disadvantage. There is a limit to the size of DNA fragment that can be cloned with an M13 vector, with 1500 bp generally being looked on as the maximum capacity, though fragments up to 3 kb have occasionally been cloned. To get around this problem a number of novel vectors ('**phagemids**') have been developed by combining a part of the M13 genome with plasmid DNA. An example is provided by pEMBL8 (Figure 6.12(a)), which was made by transferring into pUC8 a 1300 bp fragment of the M13 genome. This piece of M13 DNA contains the signal sequence recognized by the enzymes that convert the normal double-stranded M13 molecule into single-stranded DNA before secretion of new phage particles. This signal sequence is still functional even though detached from the rest of the M13 genome, so pEMBL8 molecules are also converted into single-stranded DNA and secreted as defective phage particles (Figure 6.12(b)). All that is necessary is that the *E. coli* cells used as hosts for a pEMBL8 cloning experiment are subsequently infected with normal M13 to act as a **helper phage**, providing the necessary replicative enzymes and phage coat proteins. pEMBL8, being derived from pUC8, has the polylinker cloning sites within the *lacZ'* gene, so recombinant plaques can be identified in the standard way on agar containing X-gal. With pEMBL8, single-stranded versions of cloned DNA fragments up to 10 kb in length can be obtained, greatly extending the range of the M13 cloning system.

6.3 CLONING VECTORS BASED ON λ BACTERIOPHAGE

Two problems had to be solved before λ-based cloning vectors could be developed:

1. The λ DNA molecule can be increased in size by only about 5%, representing the addition of only 3 kb of new DNA. If the total size of the molecule is more than 52 kb, then it will not package into the λ head structure and infective phage particles will not

Figure 6.12 pEMBL8: a hybrid plasmid–M13 vector that can be converted into single-stranded DNA.

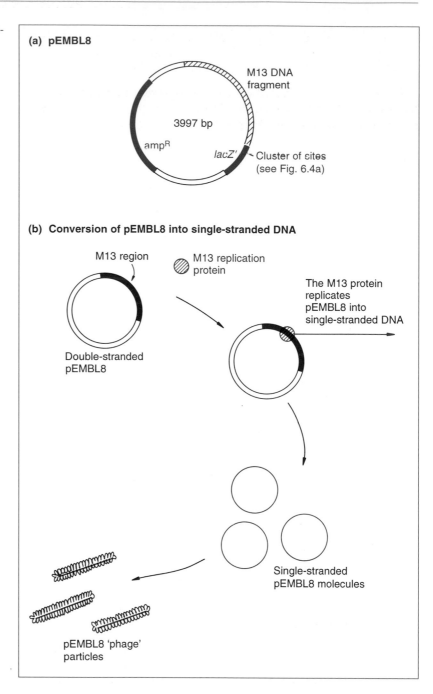

(a) pEMBL8

M13 DNA fragment

3997 bp

amp^R

lacZ'

Cluster of sites (see Fig. 6.4a)

(b) Conversion of pEMBL8 into single-stranded DNA

M13 region

M13 replication protein

The M13 protein replicates pEMBL8 into single-stranded DNA

Double-stranded pEMBL8

Single-stranded pEMBL8 molecules

pEMBL8 'phage' particles

be formed. This severely limits the size of a DNA fragment that can be inserted into an unmodified λ vector (Figure 6.13(a)).

2. The λ genome is so large that it has more than one recognition sequence for virtually every restriction endonuclease. Restriction cannot be used to cleave the normal λ molecule in a way that

will allow insertion of new DNA, as the molecule would be cut into several small fragments that would be very unlikely to reform a viable λ genome on religation (Figure 6.13(b)).

In view of these difficulties it is perhaps surprising that a wide variety of λ cloning vectors have been developed, their primary use being to clone large pieces of DNA, from 5 kb to 25 kb, much too big to be handled by plasmid or M13 vectors.

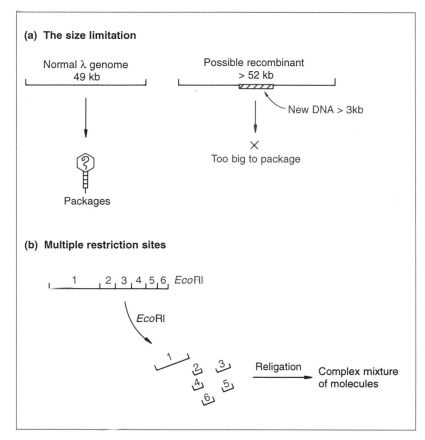

(a) The size limitation

Normal λ genome
49 kb

Possible recombinant
> 52 kb

New DNA > 3kb

Packages

✕
Too big to package

(b) Multiple restriction sites

1   2, 3, 4, 5, 6, *Eco*RI

*Eco*RI

1
2 3
4 5
6

Religation → Complex mixture of molecules

Figure 6.13 The two problems that had to be solved before λ cloning vectors could be developed. (a) The size limitation placed on the λ genome by the need to package it into the phage head. (b) λ DNA has multiple recognition sites for almost all restriction endonucleases.

6.3.1 Segments of the λ genome can be deleted without impairing viability

The way forward for the development of λ cloning vectors was provided by the discovery that a large segment in the central region of the λ DNA molecule can be removed without affecting the ability of the phage to infect *E. coli* cells. Removal of all, or part, of this non-essential region, between positions 20 and 35 on the map shown in Figure 2.9, decreases the size of the resulting λ molecule by up to 15 kb. This means that as much as 18 kb of new DNA can now be added before the cut-off point for packaging is reached (Figure 6.14).

Figure 6.14 The λ genetic map, showing the position of the non-essential region that can be deleted without affecting the ability of the phage to follow the lytic infection cycle.

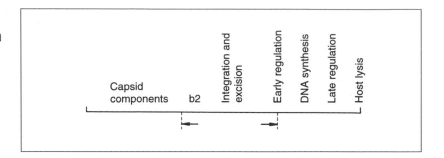

The 'non-essential' region in fact contains most of the genes involved in integration and excision of the λ prophage from the *E. coli* chromosome. A deleted λ genome is therefore non-lysogenic and can follow only the lytic infection cycle. This in itself is desirable for a cloning vector as it means induction is not needed before plaques are formed (p. 45).

6.3.2 Natural selection can be used to isolate modified λ that lack certain restriction sites

Even a deleted λ genome, with the non-essential region removed, has multiple recognition sites for most restriction endonucleases. This is a problem that is often encountered when a new vector is being developed. If just one or two sites need to be removed then the technique of *in vitro* mutagenesis (p. 221) can be used. For example, an *Eco*RI site, GAATTC, could be changed to GGATTC, which is not recognized by the enzyme. However, *in vitro* mutagenesis was in its infancy when the first λ vectors were under development, and even today would not be an efficient means of changing more than a few sites in a single molecule.

Instead, natural selection was used to provide strains of λ that lack the unwanted restriction sites. Natural selection can be brought into play by using as a host an *E. coli* strain that produces *Eco*RI. Most λ DNA molecules that invade the cell will be destroyed by the restriction endonuclease; however, a few will survive and produce plaques. These will be mutant phage, from which one or more *Eco*RI sites have been lost spontaneously (Figure 6.15). Several cycles of infection will eventually result in λ molecules that lack all or most of the *Eco*RI sites.

6.3.3 Insertion and replacement vectors

Once the problems posed by packaging constraints and by the multiple restriction sites had been solved, the way was open for the development of different types of λ-based cloning vectors. The

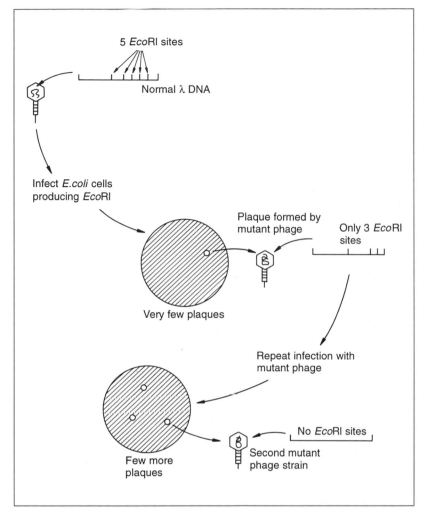

Figure 6.15 Using natural selection to isolate λ phage lacking *Eco*RI restriction sites.

first two classes of vector to be produced were λ-**insertion** and λ-**replacement** (or substitution) vectors.

(a) Insertion vectors With an insertion vector (Figure 6.16(a)), a large segment of the non-essential region has been deleted, and the two arms ligated together. An insertion vector possesses at least one unique restriction site into which new DNA can be inserted. The size of the DNA fragment that an individual vector can carry depends of course on the extent to which the non-essential region has been deleted. Two popular insertion vectors are:

- λ**gt10** (Figure 6.16(b)), which can carry up to 8 kb of new DNA, inserted into a unique *Eco*RI site located in the *c*I gene. Insertional inactivation of this gene means that recombinants are distinguished as clear rather than turbid plaques (see p. 102).

Figure 6.16 λ insertion vectors. P, polylinker in the *lacZ'* gene of λZAPII, containing unique restriction sites for *Sac*I, *Not*I, *Xba*I, *Spe*I, *Eco*RI and *Xho*I.

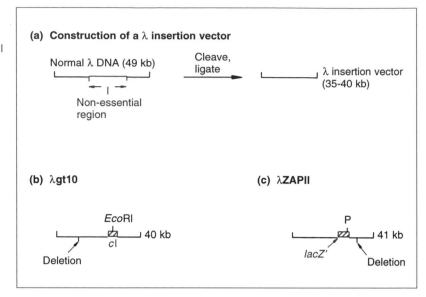

(a) **Construction of a λ insertion vector**

Normal λ DNA (49 kb)

Cleave, ligate

λ insertion vector (35–40 kb)

Non-essential region

(b) **λgt10**

*Eco*RI

cl

40 kb

Deletion

(c) **λZAPII**

P

lacZ'

41 kb

Deletion

- **λZAPII** (Figure 6.16(c)), with which insertion of up to 10 kb new DNA into any of six restriction sites within a polylinker inactivates the *lacZ'* gene carried by the vector. Recombinants give clear, rather than blue plaques on X-gal agar.

(b) Replacement vectors A λ replacement vector has two recognition sites for the restriction endonuclease used for cloning. These sites flank a segment of DNA that is replaced by the DNA to be cloned (Figure 6.17(a)). Often the replaceable fragment (or **'stuffer fragment'** in cloning jargon) carries additional restriction sites that can be used to cut it up into small pieces, so that its own reinsertion during a cloning experiment is very unlikely. Replacement vectors are generally designed to carry larger pieces of DNA than insertion vectors can handle. Recombinant selection is often on the basis of size, with non-recombinant vectors being too small to be packaged into λ phage heads (p. 103).

Two popular replacement vectors are:

- **λWES.λB'** (Figure 6.17(b)), in which two *Eco*RI sites flank the replacement fragment, and recombinant selection is solely on the basis of size.
- **λEMBL4** (Figure 6.17(c)), which can carry up to 20 kb of inserted DNA (near the theoretical maximum) by replacing a segment flanked by pairs of *Eco*RI, *Bam*HI and *Sal*I sites (either restriction endonuclease can be used to remove the stuffer fragment, so DNA fragments with a variety of sticky ends can be cloned). Recombinant selection with λEMBL4 can be on the basis of size, or can utilize the Spi phenotype (p. 102).

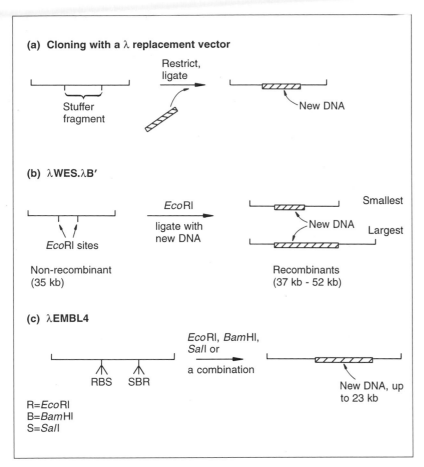

(a) Cloning with a λ replacement vector

Stuffer fragment

Restrict, ligate

New DNA

(b) λWES.λB′

EcoRI sites

EcoRI
ligate with
new DNA

Non-recombinant
(35 kb)

Smallest

New DNA

Largest

Recombinants
(37 kb - 52 kb)

(c) λEMBL4

RBS   SBR

R=*EcoRI*
B=*BamHI*
S=*SalI*

*EcoRI, BamHI,
SalI* or
a combination

New DNA, up
to 23 kb

Figure 6.17 λ replacement vectors.

6.3.4 Cloning experiments with λ insertion or replacement vectors

A cloning experiment with a λ vector can proceed along the same lines as with a plasmid vector – the λ molecules are restricted, new DNA is added, the mixture is ligated, and the resulting molecules used too transfect a competent *E. coli* host (Figure 6.18(a)). This type of experiment requires that the vector be in its circular form, with the *cos* sites hydrogen-bonded to each other.

Although quite satisfactory for many purposes, a procedure based on transfection is not particularly efficient. A greater number of recombinants will be obtained if one or two refinements are introduced. The first is to purify the two arms of the vector. When the linear form of the vector is digested with the relevant restriction endonuclease, two fragments are produced, a left arm and a right arm (Figure 6.18(b)). With virtually all λ vectors the left arm is longer than the right, and the two can be separated and purified by sucrose density gradient centrifugation. A recombinant molecule is then constructed by mixing together the DNA to be cloned with samples of the left and right arm preparations. Ligation results in several molecular arrangements, including concatamers comprising left arm–DNA–right arm repeated many times (Figure 6.18(b)). If the inserted DNA is the correct size then the *cos* sites that separate these structures will be the right distance apart for *in vitro* packaging (p. 99). Recombinant phage are therefore produced in the test-tube and can be used to infect an *E. coli* culture. This strategy, in particular the use of *in vitro* packaging, results in a large number of recombinant plaques.

Figure 6.18 Different strategies for cloning with a λ vector. (a) Using the circular form of λ as a plasmid. (b) Using purified left and right arms of the λ genome, plus *in vitro* packaging, to achieve a greater number of recombinant plaques.

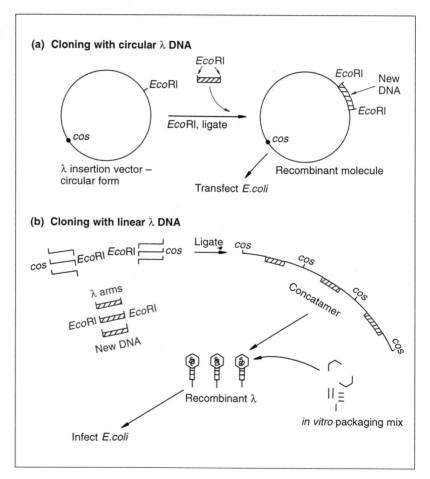

6.3.5 Very large DNA fragments can be cloned using a cosmid

The final and most sophisticated type of λ-based vector is the **cosmid**. Cosmids are hybrids between a phage DNA molecule and a bacterial plasmid, and are designed around the fact that the enzymes that package the λ DNA molecule into the phage protein coat need only the *cos* sites in order to function (p. 24). The *in vitro* packaging reaction works not only with λ genomes, but also with any molecule that carries *cos* sites separated by 37–52 kb of DNA.

A cosmid is basically a plasmid that carries a *cos* site (Figure 6.19(a)). It also needs a selectable marker, such as the ampicillin resistance gene, and a plasmid origin of replication, as cosmids lack all the λ genes and so do not produce plaques. Instead colonies are formed on selective media, just as with a plasmid vector.

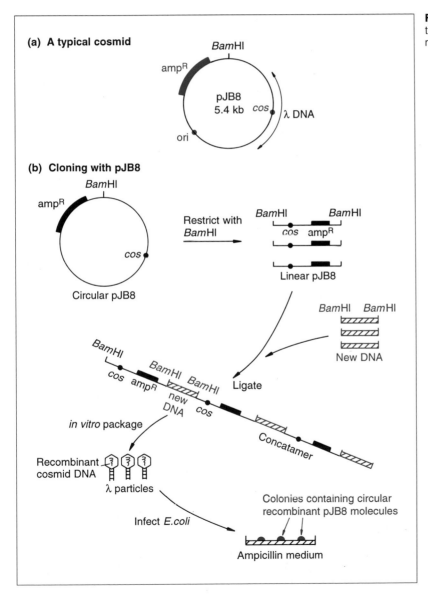

Figure 6.19 A typical cosmid and the way it is used to clone large fragments of DNA.

A cloning experiment with a cosmid is carried out as follows (Figure 6.19(b)). The cosmid is opened at its unique restriction site and new DNA fragments inserted. These fragments are produced usually by partial digestion with a restriction endonuclease, as total digestion almost invariably results in fragments that are too small to be cloned with a cosmid. Ligation is carried out so that concatamers are formed. Providing the inserted DNA is the right size, *in vitro* packaging will cleave the *cos* sites and place the recombinant cosmids in mature phage particles. These λ phage are then used to infect an *E. coli* culture, though of course plaques are not formed. Instead, infected cells are plated on to a selective medium and antibiotic-resistant colonies are grown. All colonies are recombinants as non-recombinant linear cosmids are too small to be packaged into λ heads.

6.4 λ AND OTHER HIGH CAPACITY VECTORS ENABLE GENOMIC LIBRARIES TO BE CONSTRUCTED

The main use of all λ-based vectors is to carry DNA fragments that are too large to be handled by plasmid or M13 vectors. A replacement vector such as λEMBL4 can carry up to 20 kb of new DNA, and some cosmids can manage fragments up to 40 kb. This compares with a maximum insert size of about 8 kb for most plasmids and less than 3 kb for M13 vectors.

The ability to clone such large DNA fragments raises the possibility of the **genomic library**. A genomic library is a set of recombinant clones that contain all of the DNA present in an individual organism. An *E. coli* genomic library, for example, contains all the *E. coli* genes, so any desired gene can be withdrawn from the library and studied. Genomic libraries can be retained for many years, and propagated so that copies can be sent from research group to research group.

The big question is how many clones are needed for a genomic library. The answer can be calculated with the formula:

$$N = \frac{ln(1 - P)}{ln\ (1 - a/b)}$$

where N = number of clones required
P = probability (e.g. a 95% probability that any given gene is present)
a = average size of the DNA fragment inserted into the vector
b = total size of the genome

Table 6.1   Number of clones needed for genomic libraries of a variety of organisms

Species	Genome size (bp)	Number of clones[a]	
		17 kb fragments[b]	35 kb fragments[c]
Escherichia coli	4.8×10^6	850	410
Saccharomyces cerevisiae	1.4×10^7	2 500	1 200
Drosophila melanogaster	1.7×10^8	30 000	14 500
Tomato	7.0×10^8	123 500	59 000
Man	3.0×10^9	529 000	257 000
Frog	2.3×10^{10}	4 053 000	1 969 000

[a] Calculated for a probability (P) of 95% that any particular gene will be present in the library.
[b] Fragments suitable for a replacement vector such as λEMBL4.
[c] Fragments suitable for a cosmid.

Table 6.1 shows the number of clones needed for genomic libraries of a variety of organisms, constructed using a λ replacement vector or a cosmid. It is by no means impossible to obtain several hundred thousand clones, and the methods used to identify a clone carrying a desired gene (Chapter 8) can be adapted to handle such large numbers, so genomic libraries of these sizes are by no means unreasonable. However, ways of reducing the number of clones needed for a genomic library are continually being sought. One solution is to develop new cloning vectors able to handle larger DNA inserts. During the last few years progress in this area has centred on bacteriophage **P1**, which has the advantage over λ of being able to squeeze 110 kb of DNA in to its capsid structure. Cosmid-type vectors based on P1 have been designed and used to clone DNA fragments ranging in size from 75 to 100 kb. This reduces the number of clones needed for a human genomic library from 257 000 for a λ cosmid to 90 000 for a P1 vector. Other types of novel vector, based on the F plasmid (p. 17) and called **bacterial artificial chromosomes** or **BACs**, have an even higher capacity and can handle DNA inserts up to 300 kb in size, reducing the human genomic library even further, to 30 000 clones.

6.5 VECTORS FOR OTHER BACTERIA

Cloning vectors have also been developed for several other species of bacteria, including *Streptomyces*, *Bacillus* and *Pseudomonas*. Some of these vectors are based on plasmids specific to the host organism, and some on **broad host range plasmids** able to replicate in a variety of bacterial hosts. A few are derived from bacteriophages specific to these organisms; for example, several *Streptomyces* cloning vectors are based on a λ-like phage called φC31. Most of these vectors are very similar to *E. coli* vehicles in terms of general purposes and uses and need not be considered in this book.

FURTHER READING

Bolivar, F. *et al.* (1977) Construction and characterisation of new cloning vectors. II. A multi-purpose cloning system. *Gene*, **2**, 95–113 – pBR322.

Melton, D. A., Krieg, P. A., Rebagliati, M. R., Maniatis, T., Zinn, K. and Green, M. R. (1984) Efficient *in vitro* synthesis of biologically active RNA and RNA hybridization probes from plasmids containing a bacteriophage SP6 promoter. *Nucleic Acids Research*, **12**, 7035–56 – RNA synthesis from DNA cloned in a plasmid such as pGEM3Z.

Sanger, F. *et al.* (1980) Cloning in single-stranded bacteriophage as an aid to rapid DNA sequencing. *Journal of Molecular Biology*, **143**, 161–78 – M13 vectors.

Yanisch-Perron, C., Vieira, J. and Messing, J. (1985) Improved M13 phage cloning vectors and host strains: nucleotide sequences of the M13mp18 and pUC19 vectors. *Gene*, **33**, 103–19.

Dente, L. *et al.* (1983) pEMBL: a new family of single-stranded plasmids. *Nucleic Acids Research*, **11**, 1645–55.

Leder, P. *et al.* (1977) EK2 derivatives of bacteriophage lambda useful in cloning of DNA from higher organisms: the gtWES system. *Science*, **196**, 175–7.

Frischauf, A.-M. *et al.* (1983) Lambda replacement vectors carrying polylinker sequences. *Journal of Molecular Biology*, **170**, 827–42 – the λEMBL vectors.

Collins, J. and Hohn, B. (1978) Cosmids: a type of plasmid gene cloning vector that is packageable *in vitro* in bacteriophage heads. *Proceedings of the National Academy of Sciences, USA*, **75**, 4242–6.

Sternberg, N. (1990) Bacteriophage P1 cloning system for the isolation, amplification, and recovery of DNA fragments as large as 100 kilobase pairs. *Proceedings of the National Academy of Sciences, USA*, **87**, 103–7.

Shiyuza, H., Birren, B., Kim, U. J., Mancino, V., Slepak, T., Tachiiri, Y. and Simon, M. (1992) Cloning and stable maintenance of 300-kilobase-pair fragments of human DNA in *Escherichia coli* using an F-factor-based vector. *Proceedings of the National Academy of Sciences, USA*, **89**, 8794–7.

Cloning vectors for organisms other than *E. coli*

7

Most cloning experiments are carried out with *E. coli* as the host, and the widest variety of cloning vectors are available for this organism. *E. coli* is particularly popular when the aim of the cloning experiment is to study the basic features of molecular biology such as gene structure and function. However, under some circumstances it may be desirable to use a different host for a gene cloning experiment. This is especially true in biotechnology (Chapter 12), where the aim may not be to study a gene, but to use cloning to control or improve synthesis of an important metabolic product (for example, a hormone such as insulin), or to change the properties of the organism (for example to introduce herbicide resistance into a crop plant). We must therefore consider cloning vectors for organisms other than *E. coli*.

7.1 VECTORS FOR YEAST AND OTHER FUNGI

The yeast *Saccharomyces cerevisiae* is one of the most important organisms in biotechnology. As well as its role in brewing and breadmaking, yeast has been used as a host organism for the production of important pharmaceuticals from cloned genes (p. 268). Development of cloning vectors for yeast has been stimulated greatly by the discovery of a plasmid that is present in most strains of *S. cerevisiae* (Figure 7.1). The 2 μm circle, as it is called, is one of only a very limited number of plasmids found in eukaryotic cells.

7.1.1 Selectable markers for the 2 μm plasmid

The 2 μm circle is in fact an excellent basis for a cloning vector. It is 6 kb in size, which is ideal for a vector, and exists in the yeast cell at

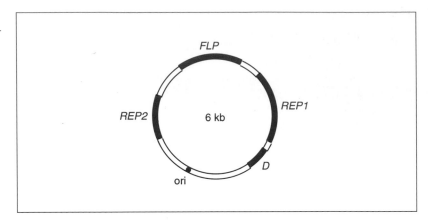

Figure 7.1 The yeast 2 µm circle. *REP1* and *REP2* are involved in replication of the plasmid, and *FLP* codes for a protein that can convert the A form of the plasmid (shown here) to the B form, in which the gene order has been rearranged by intramolecular recombination. The function of *D* is not exactly known.

a copy number of between 70 and 200. Replication makes use of a plasmid origin, several enzymes provided by the host cell, and the proteins coded by the *REP1* and *REP2* genes carried by the plasmid.

However, all is not perfectly straightforward in using the 2 µm plasmid as a cloning vector. Firstly, there is the question of a selectable marker. One or two of the most recent yeast cloning vectors carry genes conferring resistance to inhibitors such as methotrexate and copper, but most of the popular yeast vectors make use of a radically different type of selection system. In practice a normal yeast gene is used, generally one that codes for an enzyme involved in amino acid biosynthesis. An example is the gene *LEU2*, which codes for β-isopropyl malate dehydrogenase, one of the enzymes involved in the conversion of pyruvic acid to leucine.

In order to use *LEU2* as a selectable marker, a special kind of host organism is needed. The host must be an **auxotrophic** mutant that has a non-functional *LEU2* gene. Such a *leu2⁻* yeast is unable to synthesize leucine and can survive only if this amino acid is supplied as a nutrient in the growth medium (Figure 7.2(a)). Selection is possible because transformants contain a plasmid-borne copy of the *LEU2* gene, and so are able to grow in the absence of the amino acid. In a cloning experiment, cells are plated out on to **minimal medium** (which contains no added amino acids). Only transformed cells are able to survive and form colonies (Figure 7.2(b)).

7.1.2 Vectors based on the 2 µm circle – yeast episomal plasmids

Vectors derived from the 2 µm circle are called **yeast episomal plasmids** or YEps. Some YEps contain the entire 2 µm plasmid, others include just the 2µm origin of replication. An example of the latter type is YEp13 (Figure 7.3).

YEp13 illustrates several general features of yeast cloning vectors. Firstly, it is a **shuttle vector**. As well as the 2 µm origin of

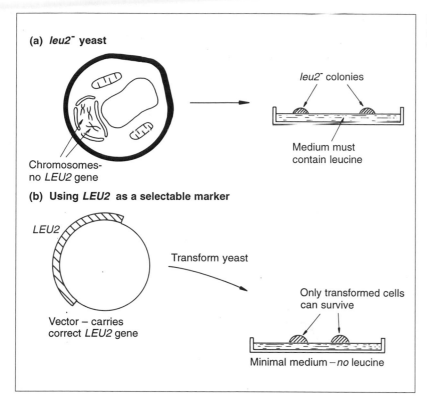

Figure 7.2 Using the *LEU2* gene as a selectable marker in a yeast cloning experiment.

replication and the selectable *LEU2* gene, YEp13 also includes the entire pBR322 sequence, and can therefore replicate and be selected for in both yeast and *E. coli*. There are several lines of reasoning behind the use of shuttle vectors. One is that it may be difficult to recover the recombinant DNA molecule from a transformed yeast colony. This is not such a problem with YEps, which are present in yeast cells primarily as plasmids, but with other yeast vectors, which may integrate into one of the yeast chromosomes (see sec-

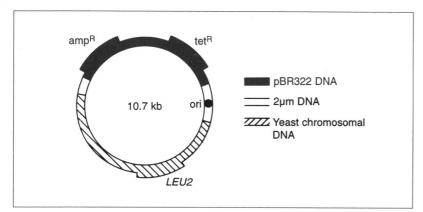

Figure 7.3 A yeast episomal plasmid – YEp13.

tion 7.1.4), purification may be impossible. This is a disadvantage as in many cloning experiments purification of recombinant DNA is essential in order for the correct construct to be identified by, for example, DNA sequencing.

The standard procedure when cloning in yeast is therefore to perform the initial cloning experiment with *E. coli*, and to select recombinants in this organism. Recombinant plasmids can then be purified, characterized, and the correct molecule introduced into yeast (Figure 7.4).

Figure 7.4 Cloning with an *E. coli*–yeast shuttle vector such as YEp13.

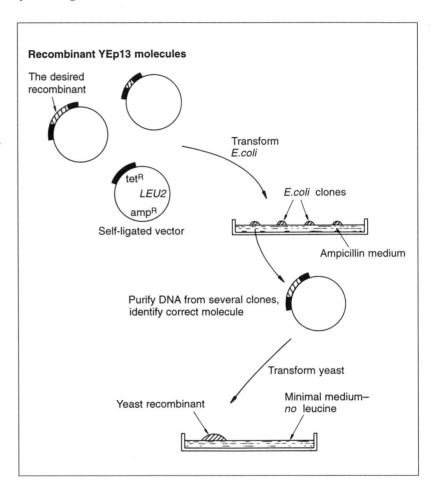

7.1.3 A YEp may insert into yeast chromosomal DNA

The word 'episomal' indicates that a YEp can replicate as an independent plasmid, but also implies that integration into one of the yeast chromosomes can occur (see the definition of 'episome' on p. 15). Integration occurs because the gene carried on the vector as a selectable marker is very closely homologous to the mutant version of the gene present in the yeast chromosomal DNA. With YEp13, for example, recombination can occur between the plasmid *LEU2* gene and the yeast mutant *LEU2* gene, resulting in insertion of the entire plasmid into one of the yeast chromosomes (Figure 7.5). The plasmid may remain integrated, or a later recombination event may result in it being excised again.

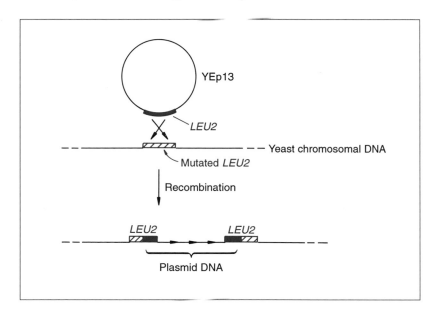

Figure 7.5 Recombination between plasmid and chromosomal *LEU2* genes can integrate YEp13 into yeast chromosomal DNA. After integration there will be two copies of the *LEU2* gene; usually one will be functional, and the other mutated.

7.1.4 Other types of yeast cloning vector

In addition to YEps, there are several other types of cloning vector for use with *S. cerevisiae*. Two important ones are as follows.

1. **Yeast integrative plasmids (YIps)**, which are basically bacterial plasmids carrying a yeast gene. An example is YIp5, which is pBR322 with an inserted *URA3* gene (Figure 7.6(a)). This gene codes for orotidine-5′-phosphate decarboxylase (an enzyme that catalyses one of the steps in the biosynthesis pathway for pyrimidine nucleotides) and is used as a selectable marker in exactly the same way as *LEU2*. A YIp can not replicate as a plasmid as it does not contain any parts of the 2 μm circle, but depends for its survival on integration into yeast chromosomal DNA. Integration occurs just as described for a YEp (Figure 7.5).
2. **Yeast replicative plasmids (YRps)**, which are able to multiply as independent plasmids because they carry a chromosomal DNA sequence that includes an origin of replication. Replication origins are known to be located very close to several yeast genes, including one or two which can be used as selectable markers. YRp7 (Figure 7.6(b)) is an example of a replicative

Figure 7.6 A YIp and a YRp.

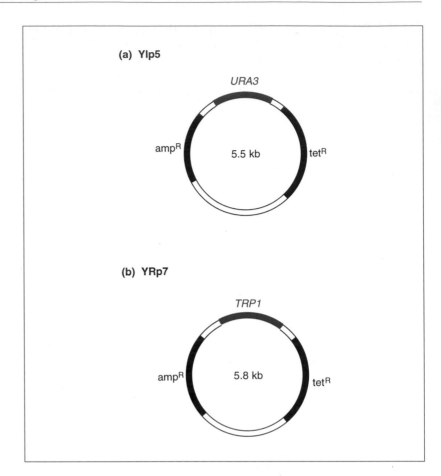

(a) **YIp5**

URA3

amp^R 5.5 kb   tet^R

(b) **YRp7**

TRP1

amp^R 5.8 kb   tet^R

plasmid. It is made up of pBR322 plus the yeast gene *TRP1*. This gene, which is involved in tryptophan biosynthesis, is located adjacent to a chromosomal origin of replication. The yeast DNA fragment present in YRp7 contains both *TRP1* and the origin.

Three factors come into play when deciding which type of yeast vector is most suitable for a particular cloning experiment. The first of these is **transformation frequency**, a measure of the number of transformants that can be obtained per μg of plasmid DNA. A high transformation frequency is necessary if a large number of recombinants are needed, or if the starting DNA is in short supply. YEps have the highest transformation frequency, providing between 10 000 and 100 000 transformed cells per μg. YRps are also quite productive, giving between 1000 and 10 000 transformants/μg, but a YIp yields less than 1000 transformants/μg, and only 1–10 unless special procedures are used. The low transformation frequency of a YIp reflects the fact that the rather rare chromosomal integration event is necessary before the vector can be retained in a yeast cell.

YEps and YRps also have the highest copy numbers, 20–50 and

5–100 respectively; in contrast a YIp is usually present at just one copy per cell. These figures are important if the objective is to obtain protein from the cloned gene, as the more copies there are of the gene the greater the expected yield of the protein product.

So why would one ever wish to use a YIp? The answer is because YIps produce very stable recombinants, as loss of a YIp that has become integrated into a chromosome occurs at only a very low frequency. YRp recombinants, on the other hand, are extremely unstable, the plasmids tending to congregate in the mother cell when a daughter buds off, so the daughter cell is non-recombinant. YEp recombinants suffer from similar problems, though an improved understanding of the biology of the 2 μm plasmid has enabled more stable YEps to be developed in recent years. Nevertheless, a YIp is the vector of choice if the needs of the experiment dictate that the recombinant yeast cells must retain the cloned gene for long periods in culture.

7.1.5 Artificial chromosomes can be used to clone huge pieces of DNA in yeast

The final type of yeast cloning vector to consider is the **YAC**, which stands for **yeast artificial chromosome**, a totally new approach to gene cloning. The development of YACs has been a spin-off from fundamental research into the structure of eukaryotic chromosomes, work that has identified the key components of a chromosome as being (Figure 7.7):

1. The centromere, which is required for the chromosome to be distributed correctly to daughter cells during cell division.
2. Two telomeres, the structures at the ends of a chromosome, which are needed in order for the ends to be replicated correctly and which also prevent the chromosome from being nibbled away by exonucleases.
3. The origins of replication, which are the positions along the chromosome at which DNA replication initiates, similar to the origin of replication of a plasmid.

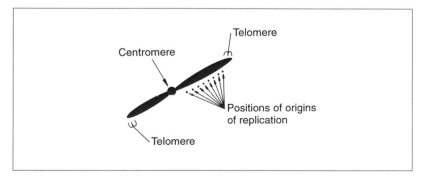

Figure 7.7 Chromosome structure.

Figure 7.8 A YAC vector and the way it is used to clone large pieces of DNA.

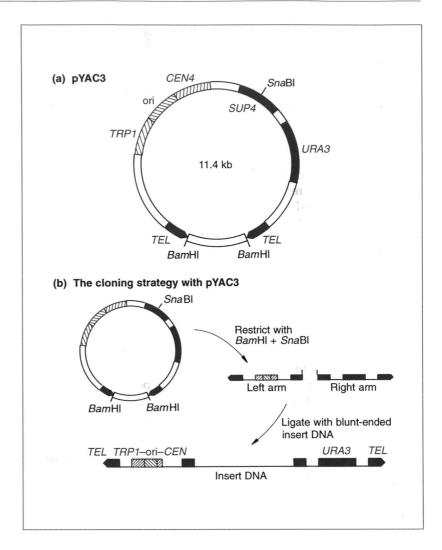

Once chromosome structure had been defined in this way the possibility arose that the individual components might be isolated by recombinant DNA techniques and then joined together again in the test-tube, creating an artificial chromosome. As the DNA molecules present in natural yeast chromosomes are several hundred kb in length, it might be possible with an artificial chromosome to clone pieces of DNA much larger than can be carried by any other type of vector.

(a) The structure and use of a YAC vector Several YAC vectors have now been developed but each one is constructed along the same lines, with pYAC3 (Figure 7.8(a)) being a typical example. At first glance pYAC3 does not look much like an artificial chromosome, but on closer examination its unique features become apparent. pYAC3 is essentially a pBR322 plasmid into which a number of yeast genes have been inserted. Two of these genes, *URA3* and *TRP1*, have been encountered already as the selectable markers for YIp5 and YRp7 respectively. As in YRp7, the DNA fragment that carries *TRP1* also contains an origin of replication, but in pYAC3 this fragment is extended even further to include the sequence called *CEN4*, which is the DNA from the centromere region of chromosome 4. The *TRP1*–origin–*CEN4* fragment therefore contains two of the three components of the artificial chromosome.

The third component, the telomeres, are provided by the two sequences called *TEL*. These are not themselves complete telomere sequences, but once inside the yeast nucleus they act as seeding sequences on to which telomeres will be built. This just leaves one other part of pYAC3 that has not been mentioned: *SUP4*, which is in fact the selectable marker into which new DNA is inserted during the cloning experiment.

The cloning strategy with pYAC3 is as follows (Figure 7.8(b)). The vector is first restricted with a combination of *Bam*HI and *Sna*BI, cutting the molecule into three fragments. The *Bam*HI fragment is removed, leaving two arms, each bounded by one *TEL* sequence and one *Sna*BI site. The DNA to be cloned, which must have blunt ends (*Sna*BI is a blunt end cutter, recognizing the sequence TACGTA), is ligated between the two arms producing the artificial chromosome. Protoplast transformation (p. 104) is then used to introduce the artificial chromosome into *S. cerevisiae*. The yeast strain that is used is a double auxotrophic mutant, *trp1⁻ ura3⁻*, which will be converted to *trp1⁺ ura3⁺* by the two markers on the artificial chromosome. Transformants are therefore selected by plating on to minimal medium, on which only cells containing a correctly constructed artificial chromosome will be able to grow. Any cells transformed with an incorrect artificial chromosome, containing two left or two right arms rather than one of each, will not be able to grow on minimal medium as one of the markers will be absent. The presence of the insert DNA in the vector can be checked by testing for insertional inactivation of *SUP4*, which is carried out by a simple colour test: white colonies are recombinants, red colonies are not.

(b) Applications for YAC vectors   The initial stimulus in designing artificial chromosomes came from yeast geneticists who wanted to use them to study various aspects of chromosome structure and behaviour, for instance to examine the segregation of chromosomes during meiosis. These experiments established that artificial chromosomes can be stably propagated in yeast cells and raised the possibility that they might be used as vehicles for genes that are too long to be cloned as a single fragment in an *E. coli* vector. Several important mammalian genes are greater than 100 kb in length (e.g. the human cystic fibrosis gene is 250 kb), beyond the capacity of all but the most sophisticated *E. coli* cloning systems (p. 131), but well within the range of a YAC vector. YACs therefore opened the way to studies of the functions and modes of expression of genes that had previously been intractable to analysis by recombinant DNA techniques. A new dimension to these experiments has recently been provided by the discovery that under some circumstances YACs can be propagated in mammalian cells, enabling the functional analysis to be carried out in the organism in which the gene normally resides.

YACs are equally important in the production of gene libraries. Recall that with fragments of 300 kb, the maximum insert size for the highest capacity *E. coli* vector, some 30 000 clones are needed for a human gene library (p. 131). YAC vectors, however, are routinely used to clone 600 kb fragments, and special types are able to handle DNA up to 1400 kb in length, the latter bringing the size of a human gene library down to just 6500 clones. Unfortunately these 'mega-YACs' have run into problems with insert stability, the cloned DNA sometimes becoming rearranged by intramolecular recombination. Nevertheless, YACs are proving to be of immense value in providing long pieces of cloned DNA, used for large-scale DNA sequencing programmes such as the Human Genome Project.

7.1.6 Vectors for other yeasts and fungi

Cloning vectors for other species of yeast and fungi are needed for basic studies of the molecular biology of these organisms and to extend the possible uses of yeasts and fungi in biotechnology. Episomal plasmids based on the *S. cerevisiae* 2 µm circle are able to replicate in a few other types of yeast, but the range of species is not broad enough for 2 µm vectors to be of general value. In any case, the requirements of biotechnology are better served by integrative plasmids, equivalent to YIps, as these provide stable recombinants that can be grown for long periods in bioreactors (p. 253). Efficient integrative vectors are now available for a number of species, including yeasts such as *Pichia pastoris* and *Kluveromyces lactis*, and the filamentous fungi *Aspergillus nidulans* and *Neurospora crassa*.

7.2 CLONING VECTORS FOR HIGHER PLANTS

There are important potential benefits of gene cloning using higher plants as the host organisms. Already gene cloning has resulted in experimental plants resistant to viruses and insects, and genetically engineered tomatoes with improved storage properties are available in supermarkets. There is real optimism that novel varieties of crops, with improved nutritional qualities and the ability to grow under adverse conditions, will be developed thanks to gene cloning.

Three types of vector system have been used with varying degrees of success with higher plants:

1. Vectors based on naturally occurring plasmids of *Agrobacterium*.
2. Direct gene transfer using DNA fragments not attached to a plant cloning vector.
3. Vectors based on plant viruses.

7.2.1 *Agrobacterium tumefaciens* – nature's smallest genetic engineer

Although no naturally occurring plasmids are known in higher plants, one bacterial plasmid, the Ti plasmid of *A. tumefaciens*, is of great importance.

A. tumefaciens is a soil microorganism that causes crown gall disease in many species of dicotyledonous plants. Crown gall occurs when a wound on the stem allows *A. tumefaciens* bacteria to invade the plant. After infection the bacteria cause a cancerous proliferation of the stem tissue in the region of the crown (Figure 7.9).

The ability to cause crown gall disease is associated with the presence of the Ti (Tumour Inducing) plasmid within the bacterial cell. This is a large (greater than 200 kb) plasmid that carries numerous genes involved in the infective process (Figure 7.10(a)). A remarkable feature of the Ti plasmid is that, after infection, part of the molecule is integrated into the plant chromosomal DNA Figure 7.10(b)). This segment, called the **T-DNA**, is between 15 and 30 kb in size, depending on the strain. It is maintained in a stable form in the plant cell and is passed on to daughter cells as an integral part of the chromosomes. But the most remarkable feature of the Ti plasmid is that the T-DNA contains eight or so genes that are expressed in the plant cell and are responsible for the cancerous properties of the transformed cells. These genes also direct synthesis of unusual compounds, called opines, that the bacteria use as nutrients (Figure 7.10(c)). In short, *A. tumefaciens* genetically engineers the plant cell for its own purposes.

(a) Using the Ti plasmid to introduce new genes into a plant cell   It was realized very quickly that the Ti plasmid could be used to transport new genes into plant cells. All that would be necessary would be to insert the new genes into the T-DNA and then the bacterium could do the hard work of integrating them into the plant chromosomal DNA. In practice this has proved quite a tricky proposition, mainly because the large size of the Ti plasmid makes manipulation of the molecule very difficult.

Figure 7.9 Crown gall disease.

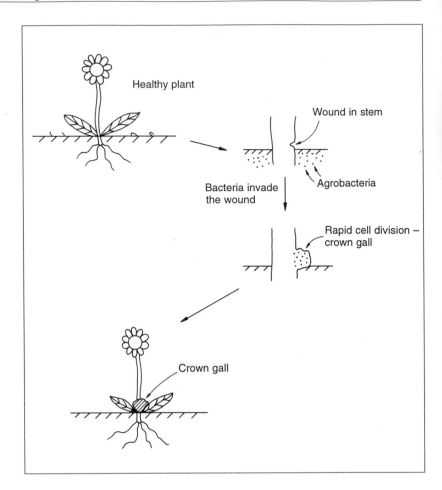

The main problem is of course that a unique restriction site is an impossibility with a plasmid 200 kb in size. Novel strategies have to be developed for inserting new DNA into the plasmid. Two are in general use.

1. **The binary vector strategy** (Figure 7.11) is based on the observation that the T-DNA does not need to be physically attached to the rest of the Ti plasmid. A two-plasmid system, with the T-DNA on a relatively small molecule, and the rest of the plasmid in normal form, is just as effective at transforming plant cells. In fact some strains of *A. tumefaciens*, and related Agrobacteria, have natural binary plasmid systems. The T-DNA plasmid is small enough to have a unique restriction site and to be manipulated using standard techniques.
2. **The cointegration strategy** (Figure 7.12) uses an entirely new plasmid, based on pBR322 or a similar *E. coli* vector, but carrying a small portion of the T-DNA. The homology between the new molecule and the Ti plasmid means that if both are present

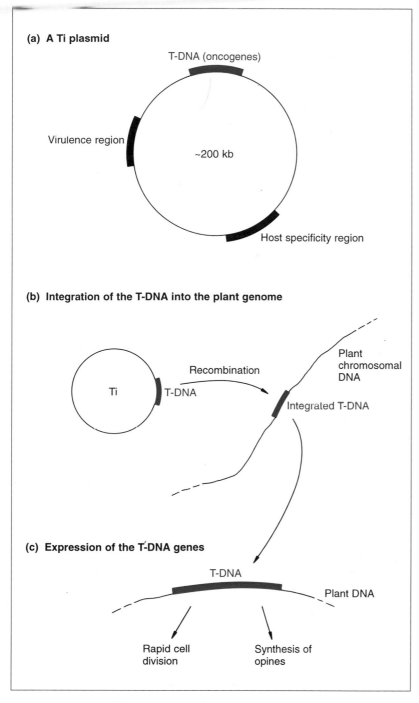

(a) A Ti plasmid

T-DNA (oncogenes)

Virulence region

~200 kb

Host specificity region

(b) Integration of the T-DNA into the plant genome

Recombination

Plant chromosomal DNA

Ti

T-DNA

Integrated T-DNA

(c) Expression of the T-DNA genes

T-DNA

Plant DNA

Rapid cell division

Synthesis of opines

Figure 7.10 The Ti plasmid and its integration into the plant chromosomal DNA after *Agrobacterium tumefaciens* infection.

in the same *A. tumefaciens* cell, then recombination can integrate the pBR plasmid into the T-DNA region. The gene to be cloned is therefore inserted into a unique restriction site on the small

Figure 7.11 The binary vector strategy. Plasmids A and B complement each other when present together in the same *Agrobacterium tumefaciens* cell. The T-DNA carried by plasmid B is transferred to the plant chromosomal DNA by proteins coded by genes carried by plasmid A.

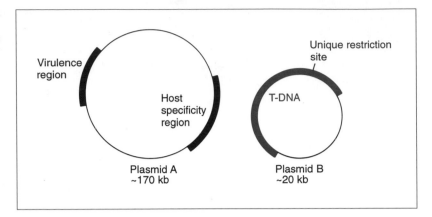

Figure 7.12 The cointegration strategy.

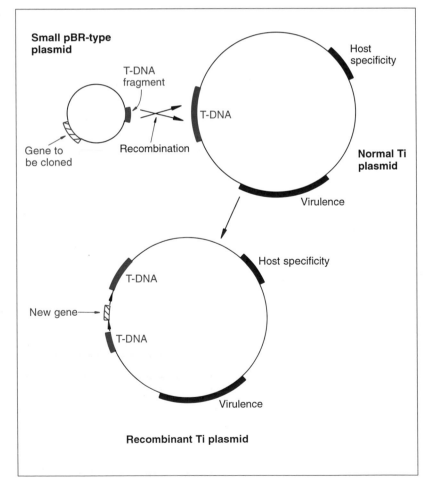

pBR plasmid, introduced into *A. tumefaciens* cells carrying a Ti plasmid, and the natural recombination process left to integrate the new gene into the T-DNA. Infection of the plant leads to insertion of the new gene, along with the rest of the T-DNA, into the plant chromosomes.

(b) Production of transformed plants with the Ti plasmid If *A. tumefaciens* bacteria that contain an engineered Ti plasmid are introduced into a plant in the natural way, by infection of a wound in the stem, then only the cells in the resulting crown gall will possess the cloned gene (Figure 7.13(a)). This is obviously of little value to the biotechnologist. Instead a way of introducing the new gene into every cell in the plant is needed.

There are several solutions, the simplest being to infect not the mature plant but a culture of plant cells or protoplasts (p. 104) in liquid medium (Figure 7.13(b)). Plant cells and protoplasts whose cell walls have reformed can be treated in the same way as microorganisms, for example they can be plated on to a selective medium in order to isolate transformants. A mature plant regenerated from transformed cells will contain the cloned gene in every cell and will pass the cloned gene to its offspring. However, regeneration of a transformed plant will occur only if the Ti vector has been '**disarmed**' so that the transformed cells do not display cancerous properties. Disarming is possible because the cancer genes, all of which lie in the T-DNA, are not needed for the infection process, infectivity being controlled mainly by the virulence region of the Ti plasmid. In fact the only parts of the T-DNA that are involved in infection are two 25 bp repeat sequences found at the left and right borders of the region integrated into the plant DNA. Any DNA placed between these two repeat sequences will be treated as 'T-DNA' and transferred to the plant. It is therefore possible to remove all the cancer genes from the normal T-DNA, and replace them with an entirely new set of genes, without disturbing the infection process.

A number of disarmed Ti cloning vectors are now available, a typical example being the binary vector pBIN19 (Figure 7.14). The left and right T-DNA borders present in this vector flank a copy of the *lacZ'* gene, containing a number of cloning sites, and a kanamycin resistance gene that functions after integration of the vector sequences into the plant chromosome. As with a yeast shuttle vector (p. 134), the initial manipulations that result in insertion of the gene to be cloned into pBIN19 are carried out in *E. coli*, the correct recombinant pBIN19 molecule then being transferred to *A. tumefaciens* and thence into the plant. Transformed plant cells are selected by plating onto agar medium containing kanamycin.

Ti vectors such as pBIN19 have recently been supplemented by related vectors based on the **Ri plasmid** of *Agrobacterium*

Figure 7.13 Transformation of plant cells by recombinant *Agrobacterium tumefaciens*. (a) Infection of a wound: transformed plant cells are present only in the crown gall. (b) Transformation of a cell suspension: all the cells in the resulting plant are transformed.

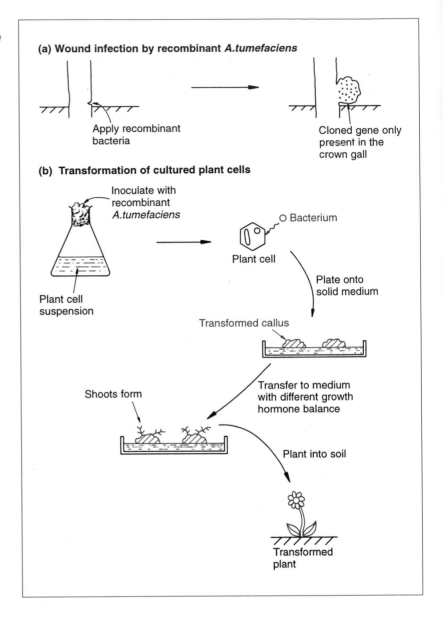

(a) Wound infection by recombinant *A.tumefaciens*

Apply recombinant bacteria

Cloned gene only present in the crown gall

(b) Transformation of cultured plant cells

Inoculate with recombinant *A.tumefaciens*

O Bacterium

Plant cell

Plant cell suspension

Plate onto solid medium

Transformed callus

Transfer to medium with different growth hormone balance

Shoots form

Plant into soil

Transformed plant

rhizogenes. Ri and Ti plasmids are very similar, the main difference being that transfer of the T-DNA from an Ri plasmid to a plant results not in a crown gall but in hairy root disease, typified by a massive proliferation of a highly branched root system. The possibility of growing transformed roots at high density in liquid culture is being explored by biotechnologists as a potential means of obtaining large amounts of protein from genes cloned in plants.

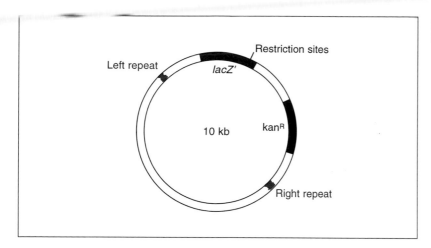

Figure 7.14 The binary Ti vector pBIN19.

7.2.2 Cloning genes in plants by direct gene transfer

Direct gene transfer is based on the observation, first made in 1984, that a supercoiled bacterial plasmid, although unable to replicate in a plant cell on its own, may be integrated by recombination into one of the plant chromosomes. The recombination event is poorly understood but is almost certainly distinct from the processes responsible for T-DNA integration. It is also distinct from the chromosomal integration of a yeast vector (p. 136), as there is no requirement for a region of homology between the bacterial plasmid and the plant DNA. In fact integration appears to occur randomly at any position in any of the plant chromosomes (Figure 7.15).

How can supercoiled plasmid DNA, carrying a selectable marker and a gene to be cloned, be introduced into a plant cell? The original methods made use of protoplasts resuspended in a viscous solution of polyethylene glycol, a polymeric negatively-charged compound that is thought to precipitate DNA on to the surfaces of the protoplasts and to induce uptake by endocytosis (Figure 7.16(a)). More efficient methods that have recently been introduced include electroporation (p. 104), microinjection (p. 104) and fusion with DNA-containing liposomes (Figure 7.16(b)). After treatment, the protoplasts are left for a few days in a solution that encourages regeneration of the cell walls, and then plated out on selective medium to identify transformants and provide callus cultures from which intact plants can be grown (exactly as described for the *Agrobacterium* system – see Figure 7.13(b)).

7.2.3 Attempts to use plant viruses as cloning vectors

Modified versions of λ and M13 bacteriophages are important cloning vectors for *E. coli* (Chapter 6), and many animal vectors are

Figure 7.15 Direct gene transfer.

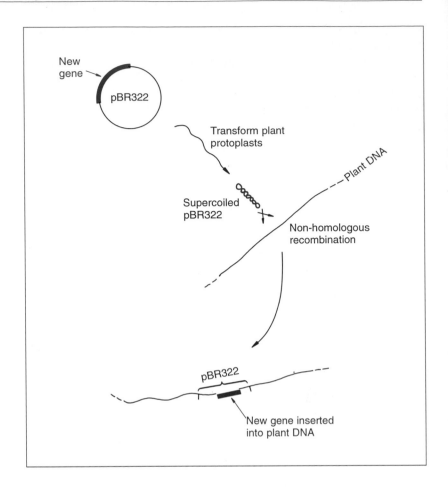

based on viruses (p. 153). Most plants are subject to viral infection, so could viruses be used to clone genes in plants? The possibility has been explored for several years but without great success. One problem is that the vast majority of plant viruses have genomes not of DNA but of RNA. RNA viruses are not so useful as potential cloning vectors because manipulations with RNA are rather more difficult to carry out. Only two classes of DNA virus are known to infect higher plants, the **caulimoviruses** and **geminiviruses**, and neither is ideally suited for gene cloning.

Although a caulimovirus vector has been used to clone a new gene into turnip plants, two general difficulties with these viruses seem certain to limit their usefulness. The first is that the total size of a caulimovirus genome is, like λ, constrained by the need to package it into its protein coat. Even after deletion of non-essential sections of the virus genome the capacity for carrying inserted DNA is still very limited, so a caulimovirus vector could only be used to clone very short pieces of DNA. This problem is compounded by the extremely narrow host range of caulimoviruses, which restricts cloning experiments to just a few plants, mainly brassicas such turnips, cabbages and cauliflowers.

   What of the geminiviruses? At first glance, these appear more promising as they naturally infect important crops such as maize and wheat. However, during the infection cycle the geminivirus genome undergoes rearrangements and deletions, which could scramble up any additional DNA that had been inserted, an

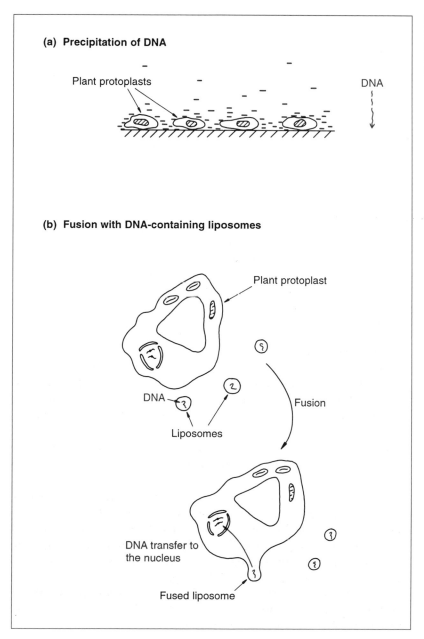

(a) Precipitation of DNA

Plant protoplasts

DNA

(b) Fusion with DNA-containing liposomes

Plant protoplast

DNA

Liposomes

Fusion

DNA transfer to
the nucleus

Fused liposome

Figure 7.16 Direct gene transfer by (a) precipitation of DNA on to the surfaces of protoplasts, and (b) fusion with DNA-containing liposomes.

obvious disadvantage for a cloning vector. There is also the problem that many of these viruses cause damaging infections to crops, so the use of a geminivirus cloning vector would be subject to stringent controls, to avoid the danger of the vector escaping from the host and infecting natural plant populations. It is therefore unlikely that either geminiviruses or caulimoviruses will ever routinely be used as plant cloning vectors, except possibly for a few specialized applications.

7.2.4 Problems with cloning genes in monocotyledonous plants

Higher plants are divided into two broad categories, the monocots and the dicots. Several factors have combined to make it much easier to clone genes in dicots such as tomato, tobacco, potato, peas and beans, but much more difficult to obtain the same results with monocots. This has been frustrating as monocots include a number of important crops (wheat, barley, rice, maize) – plants that are the most desirable targets for genetic engineering projects.

The first difficulty stems from the fact that in nature *A. tumefaciens* and *A. rhizogenes* infect only dicotyledonous plants: monocots are outside of the normal host range. For some time it was thought that this natural barrier was insurmountable and that monocots were totally resistant to transformation with Ti and Ri vectors, but eventually artificial techniques for achieving T-DNA transfer were devised. Unfortunately this has not been the end of the story. Transformation with an *Agrobacterium* vector involves regeneration of an intact plant from a transformed protoplast, cell or callus culture. The ease with which a plant can be regenerated depends very much on the particular species involved and, once again, the most difficult plants are the monocots. Failure to develop efficient procedures for regenerating intact plants from transformed cells has meant that, despite solving the problems with T-DNA transfer, *Agrobacterium* vectors are only of limited value for gene cloning with monocots.

Direct gene transfer provides a partial but as yet incomplete solution. The recombination event that integrates supercoiled plasmid DNA into a plant chromosome appears to be equally efficient with both monocots and dicots, so direct gene transfer is an effective way of obtaining transformed cells of monocots. But with the standard methods for direct gene transfer the initial transformant is a protoplast, so the difficulty in regenerating an intact plant still remains. Attempts to circumvent this problem have centred on the use of bombardment by microprojectiles (p. 104) to introduce plasmid DNA directly into plant embryos. Although this is a fairly violent transformation procedure it does not appear to be too damaging for the embryos, which still continue their normal devel opment programme to produce mature plants. The approach has been successful with maize and several other important monocots, and holds a great deal of promise for the future.

7.3 CLONING VECTORS FOR ANIMAL CELLS

Considerable effort has been put into the development of vector systems for cloning genes in animal cells. These vectors are needed in biotechnology for the synthesis of proteins from genes that do not function correctly when cloned in *E. coli* or yeast (Chapter 12), and are also being sought by clinical molecular biologists attempting to devise methodology for **gene therapy** (p. 291), the correction of inherited genetic defects by replacement of a mutant gene with a normal gene introduced into the patient by cloning techniques.

Direct gene transfer is possible with animal cells, but is not very efficient. Mammalian artificial chromosomes, similar in concept to YACs (p. 139), are also a possibility but a workable system has not yet been reported. Most cloning experiments with animal cells therefore depend on vectors based on viruses.

7.3.1 Vectors based on animal viruses

The first cloning experiment involving animal cells was carried out in 1979 with a vector based on simian virus 40 (SV40). This virus is capable of infecting several mammalian species, following a lytic cycle in some hosts and a lysogenic cycle in others. The genome is 5.2 kb in size (Figure 7.17(a)) and contains two sets of genes, the 'early' genes, expressed early in the infection cycle and coding for proteins involved in viral DNA replication, and the 'late' genes, coding for viral capsid proteins. SV40 suffers from the same problem as λ and the plant caulimoviruses, in that packaging constraints limit the amount of new DNA that can be inserted into the genome. Cloning with SV40 therefore involves replacing one or more of the existing genes with the DNA to be cloned. In the original experiment a segment of the late gene region was replaced (Figure 7.17(b)), but early gene replacement is also an option.

Since 1979 a number of other types of virus have been used to clone genes in animals. **Adenoviruses** were introduced because they enable larger fragments of DNA to be cloned than is possible with an SV40 vector, though they are more difficult to handle because the genomes are bigger. **Papillomaviruses**, which also have a relatively high capacity for inserted DNA, have the important advantage of enabling a stable transformed cell line to be obtained. Many mammalian viruses kill their host cells soon after infection, so special tricks have to be used if these are to be used for anything other than short-term transformation experiments.

Figure 7.17 SV40 and an example of its use as a cloning vector. To clone the rabbit β-globin gene the *Hind*III to *Bam*HI restriction fragment was deleted (resulting in SVGT-5) and replaced with the rabbit gene.

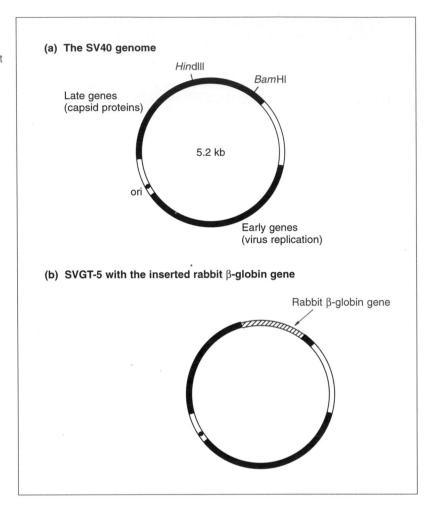

Bovine papillomavirus (BPV), which causes warts on cattle, has an unusual infection cycle in mouse cells, taking the form of a multi-copy plasmid with about 100 molecules present per cell. It does not cause the death of the mouse cell, and BPV molecules are passed to daughter cells on cell division. Shuttle vectors consisting of BPV and pBR322 sequences, and capable of replication in both mouse and bacterial cells, are therefore of great value in animal cell biotechnology. Biotechnologists are also excited by the possibilities provided by the baculoviruses, which enable large amounts of protein to be obtained from genes cloned in insect cells. These special applications of animal vectors will be discussed in Chapter 12.

FURTHER READING

Broach, J. R. (1982) The yeast 2 μm circle. *Cell*, **28**, 203–4.

Parent, S. A. *et al.* (1985) Vector systems for the expression, analysis and cloning of DNA sequences in *S. cerevisiae*. *Yeast*, **1**, 83–138 – details of different yeast cloning vectors.

Burke, D. T. *et al.* (1987) Cloning of large segments of exogenous DNA into yeast by means of artificial chromosome vectors. *Science*, **236**, 806–12.

Chilton, M. D. (1983) A vector for introducing new genes into plants. *Scientific American*, **248** (June), 50–9 – the Ti plasmid.

Bevan, M. (1984) Binary *Agrobacterium* vectors for plant transformation. *Nucleic Acids Research*, **12**, 8711–21.

Paszkowski, J. *et al.* (1984) Direct gene transfer to plants. *EMBO Journal*, **3**, 2717–22.

Brisson, N. *et al.* (1984) Expression of a bacterial gene in plants by using a viral vector. *Nature*, **310**, 511–14 – a cloning experiment with a caulimovirus.

Davies, J. W. and Stanley, J. (1989) Geminivirus genes and vectors. *Trends in Genetics*, **5**, 77–81.

Monaco, A. P. and Larin, Z. (1994) YACs, BACs, PACs and MACs: artificial chromosomes as research tools. *Trends in Biotechnology*, **12**, 280–6.

Hamer, D. H. and Leder, P. (1979) Expression of the chromosomal mouse β-maj-globin gene cloned in SV40. *Nature*, **281**, 35–40.

Graham, F. L. (1990) Adenoviruses as expression vectors and recombinant vaccines. *Trends in Biotechnology*, **8**, 20–5.

Part Two
The Applications of Cloning in Gene Analysis

How to obtain a clone of a specific gene

8

So far in this book gene cloning has been considered somewhat as an end in itself, with no attempt made to put the technique into context with molecular biological research as a whole. In particular, the examples of gene cloning used to illustrate the methodology have made use of 'DNA fragments' of unspecified origin and indeterminate importance. The astute reader will by now be wondering how construction of recombinant DNA molecules and transformation of living cells are used in the real world to obtain clones of specific genes and to provide information of importance to molecular biology and biotechnology.

In this part of the book, the role and relevance of gene cloning should become clear. Firstly, in this chapter, the methods available for obtaining a clone of an individual, specified gene will be described. This is in fact the critical test of a gene cloning experiment – success or failure often depends on whether or not a strategy can be devised by which clones of the desired gene can be selected directly, or alternatively, distinguished from other recombinants. Once this problem has been resolved, and a clone has been obtained, the molecular biologist is able to make use of a wide variety of different techniques which will extract information about the gene. The most important of these will be described in Chapters 9, 10 and 11.

8.1 THE PROBLEM OF SELECTION

The problem faced by the molecular biologist wishing to obtain a clone of a single, specified gene was illustrated in Figure 1.3. Even the simplest organisms, such as *E. coli*, contain several thousand genes, and a restriction digest of total cell DNA will produce not only the fragment carrying the desired gene, but also many other fragments carrying all the other genes (Figure 8.1(a)). During the ligation reaction there is of course no selection for an individual fragment: numerous different recombinant DNA molecules are produced, all containing different pieces of DNA (Figure 8.1(b)). Consequently a variety of recombinant clones are obtained after transformation and plating out (Figure 8.1(c)). Somehow the correct one must be identified.

Figure 8.1 The problem of selection.

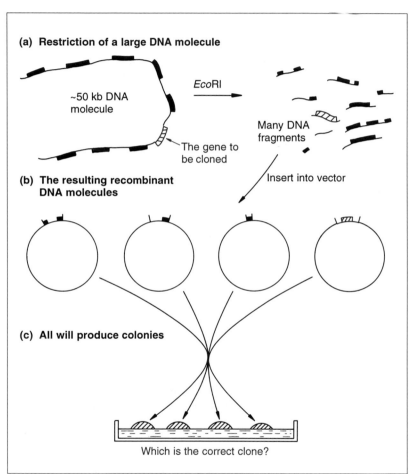

(a) **Restriction of a large DNA molecule**

~50 kb DNA molecule

*Eco*RI

The gene to be cloned

Many DNA fragments

Insert into vector

(b) **The resulting recombinant DNA molecules**

(c) **All will produce colonies**

Which is the correct clone?

8.1.1 There are two basic strategies for obtaining the clone you want

Although there are many different procedures by which the desired clone can be obtained, all are variations on two basic themes.

1. **Direct selection for the desired gene** (Figure 8.2(a)), which means that the cloning experiment is designed in such a way

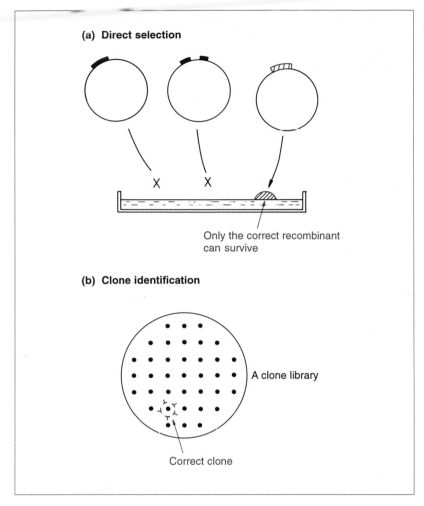

(a) Direct selection

X   X

Only the correct recombinant
can survive

(b) Clone identification

A clone library

Correct clone

Figure 8.2 The basic strategies that
can be used to obtain a particular
clone. (a) Direct selection.
(b) Identification of the desired
recombinant from a clone library.

that the only clones that are obtained are clones of the required
gene. Almost invariably, selection occurs at the plating-out
stage.

2. **Identification of the clone from a gene library** (Figure 8.2(b)),
which entails an initial **'shotgun'** cloning experiment, to produce a clone library representing all or most of the genes present in the cell, followed by analysis of the individual clones to
identify the correct one.

In general terms, direct selection is the preferred method, as it is
quick and usually unambiguous. However, as we shall see, it is not
applicable to all genes. Techniques for clone identification are
therefore very important, especially as complete genomic libraries
of many organisms are now available.

8.2 DIRECT SELECTION

To be able to select for a cloned gene it is necessary to plate the transformants on to an agar medium on which only the desired recombinants, and no others, can grow. The only colonies that are obtained will therefore be ones that comprise cells containing the desired recombinant DNA molecule.

The simplest example of direct selection occurs when the desired gene specifies resistance to an antibiotic. As an example we will consider an experiment to clone the gene for kanamycin resistance from plasmid R6-5. This plasmid in fact carries genes for resistances to four antibiotics: kanamycin, chloramphenicol, streptomycin and sulphonamide. The kanamycin resistance gene lies within one of the 13 *Eco*RI fragments (Figure 8.3(a)).

To clone this gene the *Eco*RI fragments of R6-5 would be inserted into the *Eco*RI site of a vector such as pBR322. The ligated mix will comprise many copies of 13 different recombinant DNA molecules, one set of which carries the gene for kanamycin resistance (Figure 8.3(b)).

Insertional inactivation cannot be used to select recombinants when the *Eco*RI site of pBR322 is used. This is because this site does not lie in either the ampicillin or the tetracycline resistance genes of this plasmid (Figure 6.1). But this is immaterial for cloning the kanamycin resistance gene as in this case the cloned gene can be used as the selectable marker. Transformants are plated on to kanamycin agar, on which the only cells able to survive and produce colonies are those recombinants that contain the cloned kanamycin resistance gene (Figure 8.3(c)).

8.2.1 Marker rescue extends the scope of direct selection

Direct selection would be very limited indeed if it could be used only for cloning antibiotic resistance genes. Fortunately the technique can be extended by making use of mutant strains of *E. coli* as the hosts for transformation.

As an example, consider an experiment to clone the gene *trpA* from *E. coli*. This gene codes for the enzyme tryptophan synthase, which is involved in biosynthesis of the essential amino acid tryptophan. A mutant strain of *E. coli* that has a non-functional *trpA* gene is called *trpA⁻*, and is able to survive only if tryptophan is added to the growth medium. *E. coli trpA⁻* is therefore another example of an auxotroph (p. 134).

This *E. coli* mutant can be used to clone the correct version of the *trpA* gene. Total DNA is first purified from a normal (wild-type) strain of the bacterium. Digestion with a restriction endonuclease, followed by ligation into a vector, produces numerous recombi-

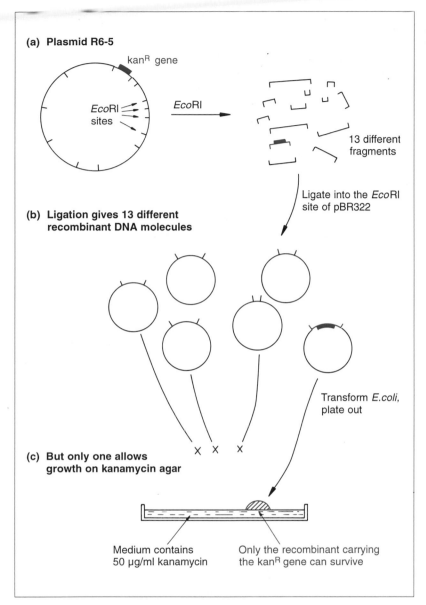

Figure 8.3 Direct selection for the cloned R6-5 kanamycin resistance (kan^R) gene.

nant DNA molecules, one of which may, with luck, carry an intact copy of the *trpA* gene (Figure 8.4(a)). This will of course be the functional gene as it has been obtained from the wild-type strain.

The ligation mixture is now used to transform the auxotrophic *E. coli trpA*⁻ cells (Figure 8.4(b)). The vast majority of the resulting transformants will be auxotrophic, but a few will now have the plasmid-borne copy of the correct *trpA* gene. These recombinants will be non-auxotrophic – they will no longer require tryptophan as the cloned gene will be able to direct production of tryptophan

Figure 8.4 Direct selection for the *trpA* gene cloned in a *trpA⁻* strain of *E. coli*.

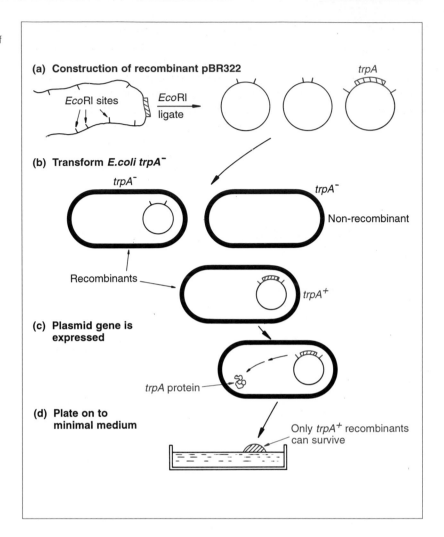

(a) **Construction of recombinant pBR322**

EcoRI sites

$\xrightarrow[\text{ligate}]{EcoRI}$

trpA

(b) **Transform E.coli trpA⁻**

trpA⁻

trpA⁻

Non-recombinant

Recombinants

trpA⁺

(c) **Plasmid gene is expressed**

trpA protein

(d) **Plate on to minimal medium**

Only *trpA⁺* recombinants can survive

synthase (Figure 8.4(c)). Direct selection is therefore performed by plating transformants on to minimal medium, which lacks any added supplements, and in particular has no tryptophan (Figure 8.4(d)). Auxotrophs can not grow on minimal medium, so the only colonies to appear will be recombinants that contain the cloned *trpA* gene.

8.2.2 The scope and limitations of marker rescue

Although marker rescue can be used to obtain clones of many genes, the technique is subject to two limitations.

1. A mutant strain must be available for the gene in question.
2. A medium on which only the wild-type can survive is needed.

In general terms, marker rescue is applicable for genes that code for biosynthetic enzymes, as these can be selected on minimal medium in the manner described for *trpA*. However, the technique is not limited to *E. coli* nor even bacteria; auxotrophic strains of yeast and filamentous fungi are also available, and marker rescue has been used to select genes cloned into these organisms.

In addition, *E. coli* auxotrophs can be used as hosts for the selection of some genes from other organisms. Often there is sufficient similarity between equivalent enzymes from different bacteria, or even from yeast, for the foreign enzyme to function in *E. coli*, so that the cloned gene is able to transform the host to wild-type.

8.3 IDENTIFICATION OF A CLONE FROM A GENE LIBRARY

Although marker rescue is a powerful technique it is not all-embracing and there are many important genes that cannot be selected by this method. Many bacterial mutants are not auxotrophs, so the mutant and wild-type strains cannot be distinguished by plating on to minimal or any other special medium. In addition, neither marker rescue nor any other direct selection method is of much use in providing bacterial clones of genes from higher organisms (i.e. animals and plants), as in these cases the differences are usually so great that the foreign enzymes do not function in the bacterial cell.

The alternative strategy must therefore be considered. This is where a large number of different clones are obtained and the desired one identified in some way.

8.3.1 Gene libraries

Before looking at the methods used to identify individual clones, the library itself must be considered. A genomic library (p. 130) is a collection of clones sufficient in number to be likely to contain every single gene present in a particular organism. Genomic libraries are prepared by purifying total cell DNA, and then making a partial restriction digest, resulting in fragments that can be cloned into a suitable vector (Figure 8.5), usually a λ replacement vector, a cosmid, or possibly a YAC.

For bacteria, yeast and fungi, the number of clones needed for a complete genomic library is not so large as to be unmanageable (Table 6.1). For plants and animals, though, a complete library contains so many different clones that identification of the desired one may prove a mammoth task. With these organisms a second type of library, specific not to the whole organism but to a particular cell type, may be more useful.

Figure 8.5 Preparation of a gene library in a cosmid vector.

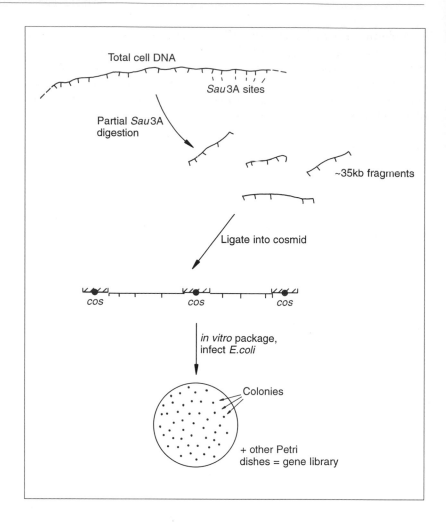

8.3.2 Not all genes are expressed at the same time

A characteristic of most multicellular organisms is specialization of individual cells. A human being, for example, is made up of a large number of different cell types – brain cells, blood cells, liver cells, etc. Each cell of course contains the same complement of genes, but in different cell types different sets of genes are switched on, others are silent (Figure 8.6).

The fact that only relatively few genes are expressed in any one type of cell can be utilized in preparation of a library if the material that is cloned is not DNA but **messenger RNA (mRNA)**. Only those genes that are being expressed are transcribed into mRNA, so if mRNA is used as the starting material then the resulting clones will comprise only a selection of the total number of genes in the cell.

A cloning method that uses mRNA would be particularly useful if the desired gene is expressed at a high rate in an individual cell type. For example, the gene for gliadin, one of the nutritionally important proteins present in wheat, is expressed at a very high level in the cells of developing wheat seeds. In these cells over 30% of the total mRNA specifies gliadin. Clearly if we could clone the mRNA from wheat seeds then we would obtain a large number of clones specific for gliadin.

8.3.3 mRNA can be cloned as complementary DNA

mRNA cannot itself be ligated into a cloning vector. However, mRNA can be converted into DNA by **complementary DNA (cDNA)** synthesis.

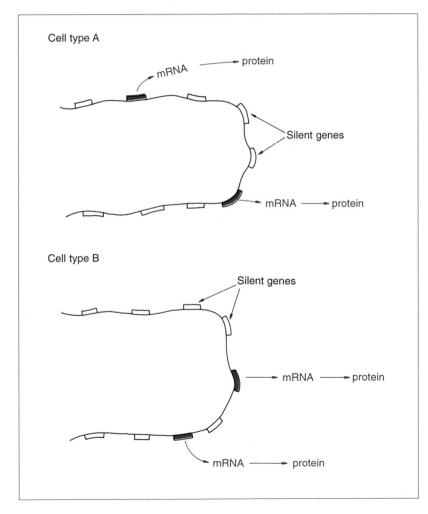

Figure 8.6 Different genes are expressed in different types of cell.

The key to this method is the enzyme reverse transcriptase (p. 58) which synthesizes a DNA polynucleotide strand complementary to an existing RNA strand (Figure 8.7(a)). Once the cDNA strand has been synthesized the RNA member of the hybrid molecule can be partially degraded by treating with ribonuclease H (Figure 8.7(b)). The remaining RNA fragments then serve as primers (p. 57) for DNA polymerase I, which synthesizes the second cDNA strand (Figure 8.7(c)), resulting in a double-stranded DNA fragment that can be ligated into a vector and cloned (Figure 8.7(d)).

The resulting cDNA clones will be representative of the mRNA present in the original preparation. In the case of mRNA prepared from wheat seeds, the cDNA library will contain a large proportion of clones representing gliadin mRNA (Figure 8.7(e)). Other clones will also be present, but locating the cloned gliadin cDNA is a much easier process than identifying the equivalent gene from a complete wheat genomic library.

8.4 METHODS FOR CLONE IDENTIFICATION

Once a suitable library has been prepared, a number of procedures can be employed to attempt identification of the desired clone. Although a few of these procedures are based on detection of the translation product of the cloned gene, it is usually easier to identify directly the correct recombinant DNA molecule. This can be achieved by the important technique of **hybridization probing**.

8.4.1 Complementary nucleic acid strands hybridize to each other

Any two single-stranded nucleic acid molecules have the potential to form base pairs with one another. With most pairs of molecules the resulting hybrid structures are unstable, as only a small number of individual interstrand bonds are formed (Figure 8.8(a)). However, if the polynucleotides are complementary then extensive base-pairing will occur to form a stable double-stranded molecule (Figure 8.8(b)). Not only can this occur between single-stranded DNA molecules to form the DNA double helix, but also between single-stranded RNA molecules and between combinations of one DNA and one RNA strand (Figure 8.8(c)).

Nucleic acid hybridization can be used to identify a particular recombinant clone if a DNA or RNA probe, complementary to the desired gene, is available. The exact nature of the probe will be discussed later in the chapter. First we must consider the technique itself.

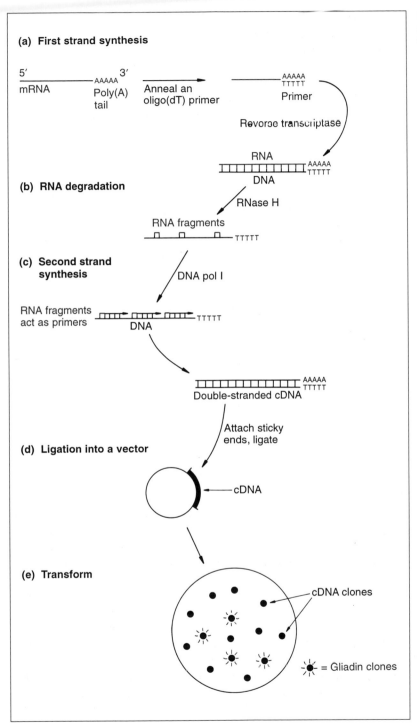

(a) First strand synthesis

(b) RNA degradation

(c) Second strand synthesis

(d) Ligation into a vector

(e) Transform

Figure 8.7 One possible scheme for cDNA cloning. See the text for details.

Figure 8.8 Nucleic acid hybridization. (a) An unstable hybrid molecule formed between two non-homologous DNA strands. (b) A stable hybrid formed between two complementary strands. (c) A DNA–RNA hybrid, such as may be formed between a gene and its transcript.

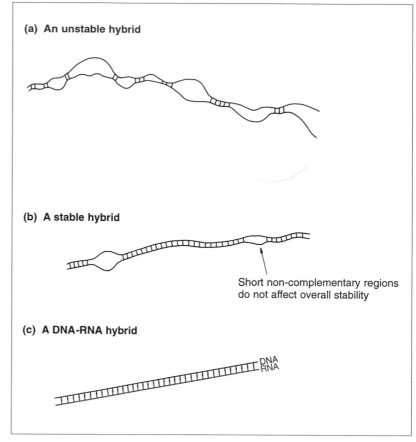

(a) **An unstable hybrid**

(b) **A stable hybrid**

Short non-complementary regions do not affect overall stability

(c) **A DNA-RNA hybrid**

DNA
RNA

8.4.2 Colony and plaque hybridization probing

Hybridization probing can be used to identify recombinant DNA molecules contained in either bacterial colonies or bacteriophage plaques. Thanks to innovative techniques developed in the late 1970s it is no longer necessary to purify each recombinant molecule. Instead an *in situ* probing method is used.

First the colonies or plaques are transferred to a nitrocellulose or nylon membrane (Figure 8.9(a)) and then treated to remove all contaminating material, leaving just DNA (Figure 8.9(b)). Usually this treatment also results in denaturation of the DNA molecules, so that the hydrogen bonds between individual strands in the double helix are broken. These single-stranded molecules can then be bound tightly to the membrane by a short period at 80°C if a nitrocellulose membrane is being used, or with a nylon membrane by ultraviolet irradiation. The molecules will in fact be attached to the membrane through their sugar–phosphate backbones, so the bases will be free to pair with complementary nucleic acid molecules.

The probe must now be labelled, denatured by heating, and

applied to the membrane in a solution of chemicals that promote nucleic acid hybridization (Figure 8.9(c)). After a period to allow hybridization to take place, the filter is washed to remove unbound probe, dried, and the positions of the bound probe detected (Figure 8.9(d)).

Traditionally the probe is labelled with a radioactive nucleotide, either by nick translation or end-filling (p. 70), or alternatively by **random priming** (Figure 8.10), a technique that results in a probe

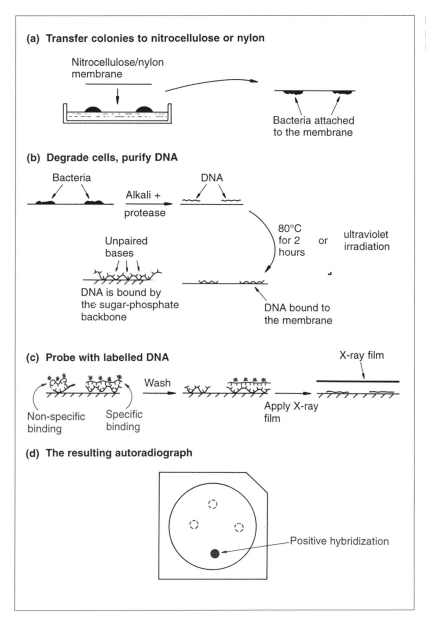

Figure 8.9 Colony hybridization probing with a radioactively labelled probe.

Figure 8.10 Labelling DNA by random priming. The mixture of random hexamers (hexameric oligonucleotides of random sequence) is sufficiently complex to include at least a few molecules that can base-pair with the probe.

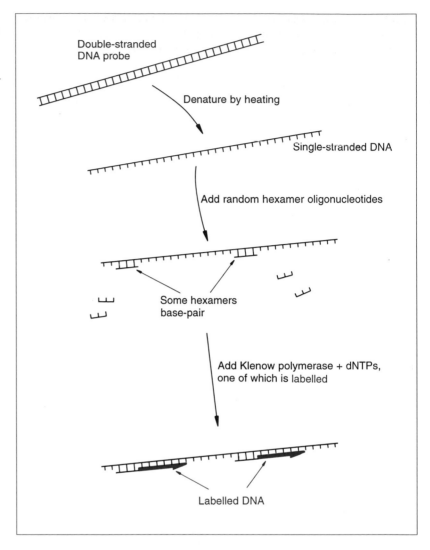

with higher activity and therefore able to detect smaller amounts of membrane-bound DNA. With these methods the position of the hybridization signal is determined by autoradiography. Radioactive labelling methods are starting to fall out of favour, however, partly because of the hazard to the researcher and partly because of the problems associated with disposal of radioactive waste. The hybridization probe may therefore be labelled in a non-radioactive manner. A number of methods have been developed, two of which are illustrated in Figure 8.11. The first makes use of dUTP nucleotides modified by reaction with **biotin**, an organic molecule that has a high affinity for a protein called **avidin**. After hybridization the positions of the bound biotinylated probe can be determined by washing with avidin coupled to a fluorescent marker

(Figure 8.11(a)). This method is as sensitive as radioactive probing and is becoming increasingly popular. The same goes for the second procedure for non-radioactive hybridization probing, where

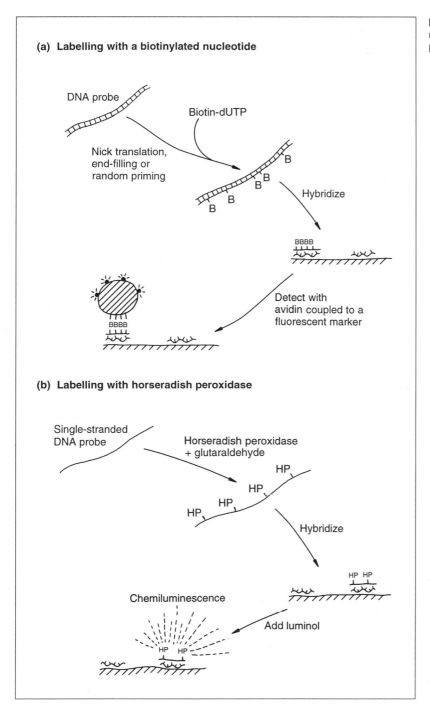

(a) **Labelling with a biotinylated nucleotide**

DNA probe

Biotin-dUTP

Nick translation, end-filling or random priming

B
B
B
B
B

Hybridize

BBBB

Detect with avidin coupled to a fluorescent marker

BBBB

(b) **Labelling with horseradish peroxidase**

Single-stranded DNA probe

Horseradish peroxidase + glutaraldehyde

HP
HP
HP
HP

Hybridize

HP HP

Chemiluminescence

Add luminol

HP HP

Figure 8.11 Two methods for the non-radioactive labelling of DNA probes.

Art Center College of Design
Library
1700 Lida Street
Pasadena, Calif. 91103

the probe DNA is complexed with the enzyme **horseradish peroxidase**, and is detected through the enzyme's ability to degrade luminol with the emission of chemiluminescence (Figure 8.11(b)). The signal can be recorded on normal photographic film in a manner analogous to autoradiography.

8.4.3 Examples of the practical use of hybridization probing

Clearly the success of colony or plaque hybridization as a means of identifying a particular recombinant clone hangs on the availability of a DNA molecule that can be used as a probe. This probe must share at least a part of the sequence of the cloned gene. If the gene itself is not available (which presumably is the case if the aim of the experiment is to provide a clone of it) then what can be used as the probe?

In practice the nature of the probe is determined by the information available about the desired gene. We will consider three possibilities.

1. Where the desired gene is expressed at a high level in a cell type from which a cDNA clone library has been prepared.
2. Where the amino acid sequence of the protein coded by the gene is completely or partially known.
3. Where the gene is a member of a family of related genes.

(a) Abundancy probing to analyse a cDNA library As described earlier in this chapter, a cDNA library is often prepared in order to obtain a clone of a gene expressed at a relatively high level in a particular cell type. In the example of a cDNA library from developing wheat seeds, a large proportion of the clones will be copies of the mRNA transcripts of the gliadin gene (Figure 8.7(e)).

Identification of the gliadin clones is simply a case of using individual cDNAs from the library to probe all the other members of the library (Figure 8.12). A clone is selected at random and the recombinant DNA molecule purified, labelled and used to probe the remaining clones. This is repeated with different clones as probes until one that hybridizes to a large proportion of the library is obtained. This abundant cDNA is considered a possible gliadin clone and analysed in greater detail (e.g. by DNA sequencing and isolation of the translation product) to confirm the identification.

(b) Oligonucleotide probes for genes whose translation products have been characterized Often the gene to be cloned will code for a protein that has already been studied in some detail. In particular the amino acid sequence of the protein may have been

apollo College
Library
Pasadena Calif 91103

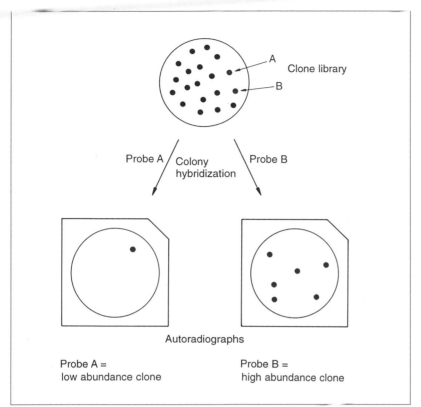

Figure 8.12 Probing within a library to identify an abundant clone.

determined, using sequencing techniques that have been available for over 30 years. If the amino acid sequence is known, then it is possible to use the genetic code to predict the nucleotide sequence of the relevant gene. This prediction is always an approximation, as only methionine and tryptophan can be assigned unambiguously to triplet codons; all other amino acids are coded by at least two codons each. Nevertheless, in most cases, the different codons for an individual amino acid are related. Alanine, for example, is coded by GCA, GCT, GCG and GCC, so two out of the three nucleotides of the triplet coding for alanine can be predicted with certainty.

As an example to clarify how these predictions are made, consider cytochrome c, a protein that plays an important role in the respiratory chain of all aerobic organisms. The cytochrome c protein from yeast was sequenced in 1963, with the result shown in Figure 8.13. This sequence contains a segment, starting at amino acid 59, that runs Trp–Asp–Glu–Asn–Asn–Met. The genetic code states that this hexapeptide is coded by TGG–GA$_C^T$–GA$_G^A$–AA$_C^T$–AA$_C^T$–ATG. Although this represents a total of 16 different possible sequences, 14 of the 18 nucleotides can be predicted with certainty.

For many years this type of prediction was of limited academic

Figure 8.13 The amino acid sequence of yeast cytochrome c. The hexapeptide that is highlighted is the one used to illustrate how a nucleotide sequence can be predicted from an amino acid sequence.

```
                                                              15
GLY—SER—ALA—LYS—LYS—GLY—ALA—THR—LEU—PHE—LYS—THR—ARG—CYS—GLU—
                                                              30
LEU—CYS—HIS—THR—VAL—GLU—LYS—GLY—GLY—PRO—HIS—LYS—VAL—GLY—PRO—
                                                              45
ASN—LEU—HIS—GLY—ILE—PHE—GLY—ARG—HIS—SER—GLY—GLN—ALA—GLN—GLY—
                                                              60
TYR—SER—TYR—THR—ASP—ALA—ASN—ILE—LYS—LYS—ASN—VAL—LEU—TRP—ASP—
                                                              75
GLU—ASN—ASN—MET—SER—GLU—TYR—LEU—THR—ASN—PRO—LYS—LYS—TYR—ILE—
                                                              90
PRO—GLY—THR—LYS—MET—ALA—PHE—GLY—GLY—LEU—LYS—LYS—GLU—LYS—ASP—
                                                              103
ARG—ASN—ASP—LEU—ILE—THR—TYR—LEU—LYS—LYS—ALA—CYS—GLU
```

interest. But nowadays short oligonucleotides of predetermined sequence can be synthesized in the laboratory (Figure 8.14). An oligonucleotide probe can therefore be constructed according to the predicted nucleotide sequence, and this probe may be able to identify the gene coding for the protein in question. In the example of yeast cytochrome c, the 16 possible oligonucleotides that can code for Trp–Asp–Glu–Asn–Asn–Met would be synthesized, either separately or as a pool, and then used to probe a yeast genomic or cDNA library (Figure 8.15). The probe will share some nucleotide homology with the cytochrome c gene, and this may be sufficient to provide an unambiguous hybridization signal. Even if more than one clone gives positive hybridization, reprobing with an oligonucleotide predicted by a different segment of the cytochrome c amino acid sequence should provide a definite result (Figure 8.15). However, the segment of the protein used for nucleotide sequence prediction must be chosen with care: the hexapeptide Ser–Glu–Tyr–Leu–Thr–Asn, which immediately follows our first choice, could be coded by several thousand different 18-nucleotide sequences, clearly an unsuitable choice for a synthetic probe.

(c) Heterologous probing allows related genes to be identified
Often a substantial amount of nucleotide **homology** is seen when two genes for the same protein, but from different organisms, are compared, a reflection of the conservation of gene structure during evolution. Frequently, two genes from related organisms are sufficiently homologous for a single-stranded probe prepared from one gene to form a stable hybrid with the second gene. Although the two molecules are not entirely complementary, enough base pairs will be formed to produce a stable structure (Figure 8.16(a)).

Heterologous probing makes use of hybridization between related sequences for clone identification. For example, the yeast cytochrome c gene, identified in the previous section by oligonucleotide probing, could itself be used as a hybridization probe to identify cytochrome c genes in clone libraries of other organisms. A

probe prepared from the yeast gene will not be entirely complementary to the gene from, say *Neurospora crassa*, but sufficient base-
pairing should occur for a hybrid to be formed and be detected by

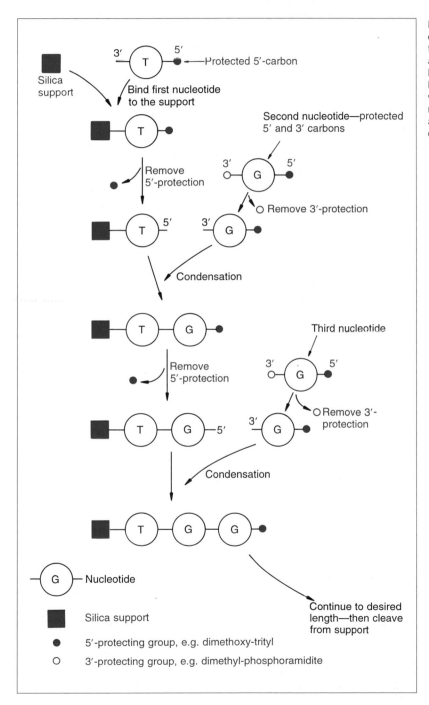

Figure 8.14 A simplified scheme for oligonucleotide synthesis. The protecting groups attached to the 3' and 5' termini prevent reactions between individual mononucleotides. By carefully controlling the time at which the protecting groups are removed, mononucleotides can be added one by one to the growing oligonucleotide.

Figure 8.15 The use of a synthetic, end-labelled oligonucleotide to identify a clone of the yeast cytochrome c gene.

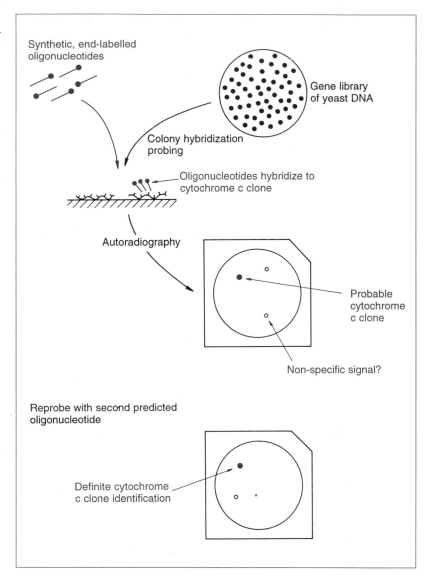

autoradiography (Figure 8.16(b)). The experimental conditions would in fact be modified so that the heterologous structure is not destabilized and lost before autoradiography.

Heterologous probing can also identify related genes in the **same** organism. If the wheat gliadin cDNA clone, identified earlier in the chapter by abundancy probing, is used to probe a genomic library, then it will hybridize not only to its own gene, but to a variety of other genes as well (Figure 8.16(c)). These are all related to the gliadin cDNA, but have slightly different nucleotide sequences. This is because the wheat gliadins form a complex group of related proteins that are coded by the members of a **multigene family**.

Once one gene in the family has been cloned, then all the other members can be isolated by heterologous probing.

8.4.4 Identification methods based on detection of the translation product of the cloned gene

Hybridization probing is usually the preferred method for identification of a particular recombinant from a clone library. The technique is easy to perform and, with modifications introduced in recent years, can be used to check up to 10 000 recombinants per experiment, allowing large genomic libraries to be screened in a reasonably short time. Nevertheless, the requirement for a probe that is at least partly complementary to the desired gene sometimes makes it impossible to use hybridization in clone identification. On these occasions a different strategy is needed.

The main alternative to hybridization probing is **immunological screening**. The distinction is that, whereas with hybridization

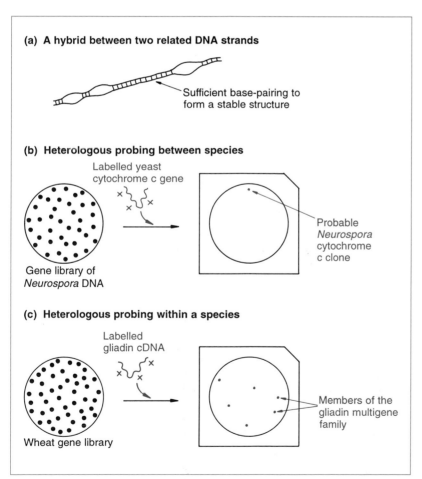

(a) **A hybrid between two related DNA strands**

Sufficient base-pairing to form a stable structure

(b) **Heterologous probing between species**

Labelled yeast cytochrome c gene

Gene library of *Neurospora* DNA

Probable *Neurospora* cytochrome c clone

(c) **Heterologous probing within a species**

Labelled gliadin cDNA

Wheat gene library

Members of the gliadin multigene family

Figure 8.16 Heterologous probing.

probing the cloned DNA fragment is itself directly identified, an immunological method detects the protein coded by the cloned gene. Immunological techniques therefore presuppose that the cloned gene is being expressed, so that the protein is being made, and that this protein is not normally present in the host cells.

(a) Antibodies are required for immunological detection methods If a purified sample of a protein is injected into the bloodstream of a rabbit, the immune system of the animal will synthesize antibodies that bind to and help degrade the foreign molecule (Figure 8.17(a)). This is of course a version of the natural defence mechanism that the animal uses to deal with invasion by bacteria, viruses and other infective agents.

Figure 8.17 Antibodies. (a) Antibodies in the bloodstream bind to foreign molecules and help degrade them. (b) Purified antibodies can be obtained from a small volume of blood taken from a rabbit injected with the foreign protein.

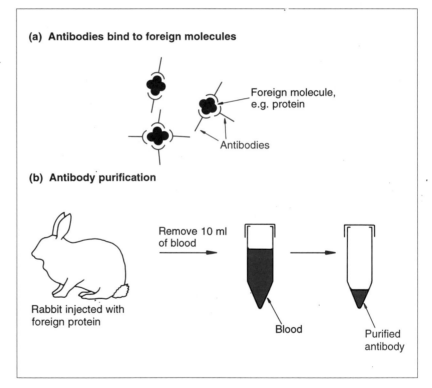

(a) Antibodies bind to foreign molecules

Foreign molecule, e.g. protein

Antibodies

(b) Antibody purification

Remove 10 ml of blood

Rabbit injected with foreign protein

Blood

Purified antibody

Once a rabbit is challenged with a protein, the levels of antibody present in its bloodstream remain high enough over the next few days for substantial quantities to be purified. It is not necessary to kill the rabbit, because as little as 10 ml of blood provides a considerable amount of antibody (Figure 8.17(b)). This purified antibody binds only to the protein with which the animal was originally challenged.

(b) Using a purified antibody to detect protein in recombinant colonies There are several versions of immunological screening, but the most useful method is a direct counterpart of colony hybridization probing. Recombinant colonies are transferred to a polyvinyl membrane, the cells are lysed, and a solution containing the specific antibody is added (Figure 8.18(a)). Either the antibody itself is labelled, or the membrane is subsequently washed with a solution of labelled **Protein A**, a bacterial protein that specifically binds to the immunoglobulins that antibodies are made of (Figure 8.18(b)). The label can be a radioactive one, in which case the colonies that bind the label are detected by autoradiography, or

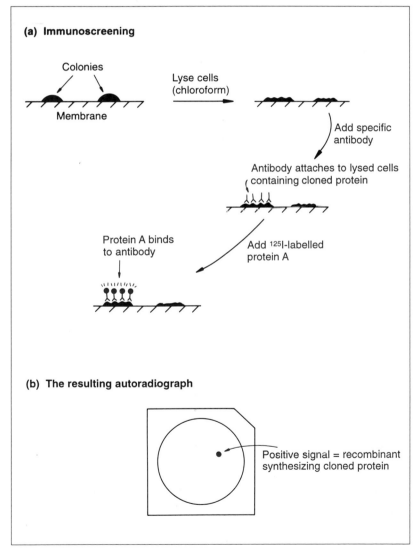

(a) **Immunoscreening**

Colonies

Membrane

Lyse cells
(chloroform)

Add specific
antibody

Antibody attaches to lysed cells
containing cloned protein

Add ^{125}I-labelled
protein A

Protein A binds
to antibody

(b) **The resulting autoradiograph**

Positive signal = recombinant
synthesizing cloned protein

Figure 8.18 Using a purified antibody to detect protein in recombinant colonies. Instead of labelled protein A, the antibody itself can be labelled, or alternatively a second labelled antibody which binds specifically to the primary antibody can be used.

non-radioactive labels resulting in a fluorescent or chemiluminescent signal can be used.

(c) The problem of gene expression   Immunological screening
obviously depends on the cloned gene being expressed so that the
protein translation product is present in the recombinant cells.
However, as will be discussed in greater detail in Chapter 12, a
gene from one organism is often not expressed in a different organism. In particular, it is very unlikely that a cloned animal or plant
gene (with the exception of chloroplast genes) will be expressed in
E. coli cells.

This problem can be circumvented by using a special type of
vector, called an **expression vector** (p. 256), designed specifically to
promote expression of the cloned gene in a bacterial host.
Immunological screening of recombinant *E. coli* colonies carrying
animal genes cloned into expression vectors has in fact been very
useful in obtaining genes for several important hormones.

FURTHER READING

Gubler, U. and Hoffman, B. J. (1983) A simple and very efficient
method for generating cDNA libraries. *Gene*, **25**, 263–9.

Dyson, N. J. (1991) Immobilization of nucleic acids and hybridization analysis, in *Essential Molecular Biology; A Practical Approach,
vol. II*, (ed. T. A. Brown), IRL Press at Oxford University Press,
Oxford, UK, pp. 111–56.

Grunstein, M. and Hogness, D. S. (1975) Colony hybridization: a
method for the isolation of cloned cDNAs that contain a specific
gene. *Proceedings of the National Academy of Sciences, USA*, **72**,
3961–5.

Benton, W. D. and Davis, R. W. (1977) Screening λgt recombinant
clones by hybridization to single plaques *in situ*. *Science*, **196**,
180–2.

Feinberg, A. P. and Vogelstein, B. (1983) A technique for labelling
DNA restriction fragments to high specific activity. *Analytical
Biochemistry*, **132**, 6–13 – random priming labelling.

Thorpe, G. H. G., Kricka, L. J., Moseley, S. B. and Whitehead, T. P.
(1985) Phenols as enhancers of the chemiluminescent horseradish peroxidase – luminol – hydrogen peroxide reaction: application in luminescence-monitored enzyme immunoassays. *Clinical
Chemistry*, **31**, 1335–41 – describes the basis to a non-radioactive
labelling method.

Young, R. A. and Davis, R. W. (1983) Efficient isolation of genes by
using antibody probes. *Proceedings of the National Academy of
Sciences, USA*, **80**, 1191 8.

Studying gene and genome structure

9

A carefully designed and skilfully performed cloning experiment will provide a colony or plaque containing copies of the recombinant DNA molecule that carries the gene of interest. In many cases, the next stage in the research project will be to purify the recombinant DNA molecule and obtain from it as much information as possible about the cloned gene.

The next three chapters cover the most important of the many techniques that have been devised for studying cloned genes. In this chapter the emphasis is placed on how to obtain information on the structure of a cloned gene and how to use cloned genes to study the structure of a genome. The relevant methods fall into three categories:

1. Techniques for determining the location of a cloned gene within a larger DNA molecule.
2. DNA sequencing methods.
3. Techniques that enable a cloned gene to be used to study the overall structure of the genome in which it normally resides.

9.1 HOW TO STUDY THE LOCATION OF A CLONED GENE

Several techniques are available for determining the location of a cloned gene on a DNA molecule. The exact nature of the procedure that is used depends on the size of the DNA molecule involved, with the techniques applicable for small molecules, such as normal and recombinant versions of plasmids and phage chromosomes, being different from those used for gene location on the large DNA molecules contained in eukaryotic chromosomes.

9.1.1 Locating the position of a cloned gene on a small DNA molecule

Consider again the example used in the previous chapter where the kanamycin resistance gene from R6-5 was cloned as an *Eco*RI fragment carried by pBR322 (p. 162). Now the clone is available it would be useful to know on which of the 13 R6-5 *Eco*RI fragments the gene lies; this information will allow the gene to be placed on the R6-5 restriction map and to be positioned relative to other genes.

First, an *Eco*RI restriction digest of R6-5 must be electrophoresed in an agarose gel so that the individual fragments can be seen (Figure 9.1(a)). One of these fragments will be the same as that inserted into the recombinant pBR322 molecule that carries the kanamycin resistance gene. The aim is therefore to label the recombinant molecule and use it to probe the restriction digest. This can be attempted while the restriction fragments are still contained in the electrophoresis gel, but the results are usually not very good, as the gel matrix causes a lot of spurious background hybridization that obscures the specific hybridization signal. Instead the DNA bands in the agarose gel are transferred to a nitrocellulose or nylon membrane, providing a much 'cleaner' environment for the hybridization experiment.

Transfer of DNA bands from an agarose gel to a membrane makes use of the technique perfected in 1975 by Professor E. M. Southern and referred to as **Southern transfer**. The membrane is placed on the gel, and buffer allowed to soak through, carrying the DNA from the gel to the membrane where the DNA is bound. Sophisticated pieces of apparatus can be purchased to assist this process, but many molecular biologists prefer a home-made set-up incorporating a lot of paper towels and considerable balancing skills (Figure 9.1(b)). The same method can also be used for the transfer of RNA molecules (**'northern' transfer**) or proteins (**'western' transfer**). So far no one has come up with 'eastern' transfers.

Southern transfer results in a membrane that carries a replica of the DNA bands from the agarose gel. If the labelled probe is now applied, hybridization will occur and autoradiography (or the equivalent detection system for a non-radioactive probe) will show which restriction fragment contains the cloned gene (Figure 9.1(c)). It will then be possible to position the kanamycin resistance gene on the R6-5 restriction map (Figure 9.1(d)).

Southern hybridization can also be used to locate the exact position of a cloned gene within a recombinant DNA molecule. This is important as often the cloned DNA fragment is relatively large (40 kb for a cosmid vector) whereas the gene of interest, contained somewhere in the cloned fragment, may be less than 1 kb in size. Also the cloned fragment may carry a number of genes in addition

Figure 9.1 Southern hybridization.

(a) Electrophorese *Eco*RI-restricted R6-5 DNA

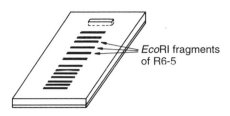

*Eco*RI fragments
of R6-5

(b) Southern transfer

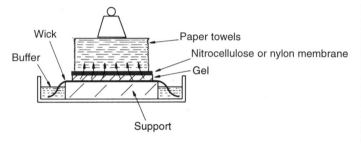

Wick

Buffer

Paper towels

Nitrocellulose or nylon membrane

Gel

Support

(c) Result of hybridization probing

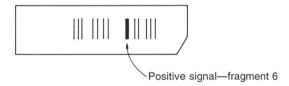

Positive signal—fragment 6

(d) Locate the fragment on the R6-5 restriction map

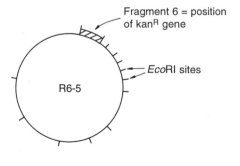

Fragment 6 = position
of kan^R gene

*Eco*RI sites

R6-5

to the one under study. The strategies described in Chapter 8 for identifying a clone from a genomic library must therefore be followed up by Southern analysis of the recombinant DNA molecule to locate the precise position within the cloned DNA fragment of the gene being sought (Figure 9.2).

Figure 9.2 Southern hybridization can be used to locate the position of a cloned gene within a recombinant DNA molecule.

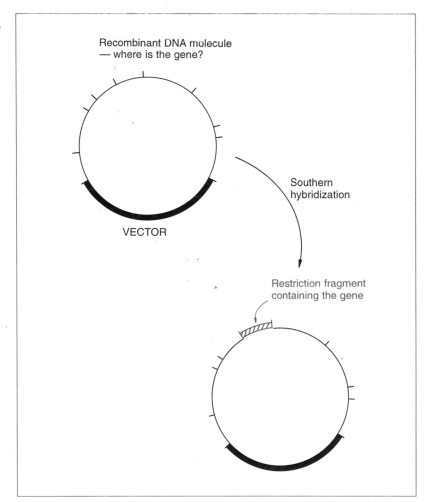

Recombinant DNA molecule
— where is the gene?

Southern hybridization

VECTOR

Restriction fragment containing the gene

9.1.2 Locating the position of a cloned gene on a large DNA molecule

Southern hybridization is feasible only if a restriction map can be worked out for the DNA molecule being studied. This means that the procedure is appropriate for most plasmids, bacteriophages and viruses, but cannot be used to locate cloned genes on larger DNA molecules. Restriction mapping becomes very complicated with molecules more than about 250 kb in size, as can be appreciated by referring back to Figure 4.18. This example of restriction mapping is relatively straightforward as the λ molecule is not very big. Imagine how much more complicated the analysis would be if there were five times as many restriction sites.

Other techniques have to be used to locate the positions of cloned eukaryotic genes on chromosomal DNA molecules. Of course if the cloned gene has previously been studied by classical genetic methods then its location will probably be known, having already been established by genetic mapping techniques. However, in many cases the most interesting cloned genes are ones that have not been identified by classical genetics. How then can such a gene be mapped?

(a) Separating chromosomes by gel electrophoresis   The first question to ask is: which chromosome carries the gene of interest? For some organisms this can be answered by a special type of Southern hybridization, involving not restriction fragments but intact chromosomal DNA molecules, separated by a novel type of gel electrophoresis.

In conventional gel electrophoresis, as described on p. 69, the electric field is orientated along the length of the gel and the DNA molecules migrate in a straight line towards the positive pole (Figure 9.3(a)). Different-sized molecules can be separated because of the different rates at which they are able to migrate through the network of pores that make up the gel. However, only molecules within a certain size range can be separated in this way, because the difference in migration rate becomes increasingly small for larger molecules (Figure 9.3(b)). In practice, molecules larger than about 50 kb cannot be resolved efficiently by standard gel electrophoresis.

The limitations of standard gel electrophoresis can be overcome if a more complex electric field is used. Several different systems have been designed but the principle is best illustrated by **orthogonal field alternation gel electrophoresis (OFAGE)**. Instead of being applied directly along the length of the gel, the electric field now alternates between two pairs of electrodes, each pair set at an angle of 45° to the length of the gel (Figure 9.4(a)). The result is a pulsed field, with the DNA molecules in the gel having continually to change direction in accordance with the pulses.

As the two fields alternate in a regular fashion the net movement of the DNA molecules in the gel is still from one end to the other, in more or less a straight line (Figure 9.4(a)). However, with every change in field direction each DNA molecule has to realign through 90° before its migration can continue. This is in fact the key point, because a short molecule can realign faster than a long one, allowing the short molecule to progress towards the bottom of the gel more quickly. This added dimension increases the resolving power of the gel quite dramatically, so that molecules up to several

Figure 9.3 Conventional agarose gel electrophoresis and its limitations.

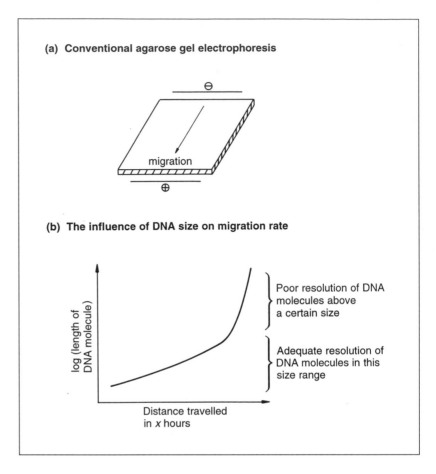

(a) Conventional agarose gel electrophoresis

migration

(b) The influence of DNA size on migration rate

log (length of DNA molecule)

Poor resolution of DNA molecules above a certain size

Adequate resolution of DNA molecules in this size range

Distance travelled in *x* hours

thousand kb can be separated. This size range includes the chromosomal molecules of many eukaryotes, including yeast, several important filamentous fungi, and protozoans such as the malaria parasite *Plasmodium falciparum*. Gels showing the chromosomes of these organisms can therefore be obtained (Figure 9.4(b)).

OFAGE and related techniques such as **CHEF (contour clamped homogeneous electric fields)** and **FIGE (field inversion gel electrophoresis)** are important for a number of reasons. For example, the DNA from individual chromosomes can be purified from the gel, enabling a series of chromosomal gene libraries to be prepared. Each of these libraries, containing the genes from just one chromosome, will be substantially smaller and easier to handle than a complete genomic library. In addition, chromosomal DNA molecules can be immobilized on a nitrocellulose or nylon membrane by Southern transfer and studied by hybridization analysis. In this way the chromosome that carries a cloned gene can be identified.

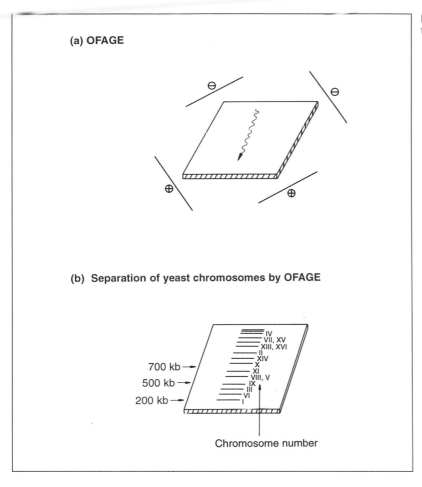

Figure 9.4 Orthogonal field alternation gel electrophoresis (OFAGE).

(b) *In situ* **hybridization to visualize the position of a cloned gene on a eukaryotic chromosome** OFAGE and other non-conventional gel electrophoresis techniques are at present limited to lower eukaryotes whose chromosomes are relatively small. The much larger molecules (>50 000 kb) of mammals and other higher eukaryotes are still some way beyond the capability of the current technology. Gene location on these larger DNA molecules can, however, be achieved by *in situ* **hybridization**, which has the added advantage of also being able to provide information regarding the position of a cloned gene on its chromosome.

In situ hybridization derives from the standard light microscopy techniques used to observe chromosomes in cells that are in the process of division (Figure 9.5(a)). With many organisms, individual chromosomes can be recognized by their shape and by the banding pattern produced by various types of stain. *In situ* hybridization provides a direct visual localization of a cloned gene on the light microscopic image of a chromosome.

Figure 9.5 Chromosomes and *in situ* hybridization.

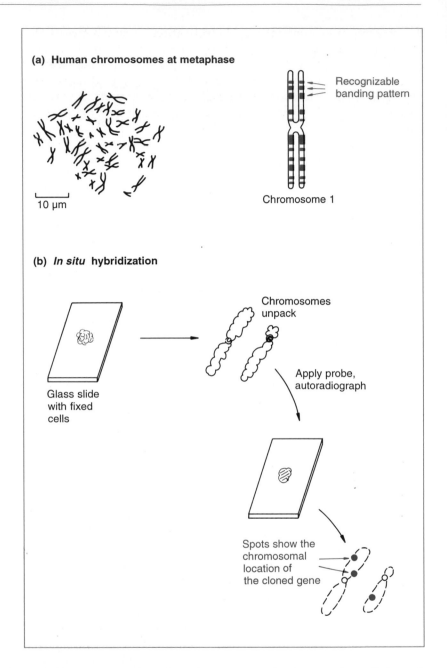

(a) **Human chromosomes at metaphase**

Recognizable
banding pattern

Chromosome 1

10 µm

(b) *In situ* hybridization

Chromosomes
unpack

Apply probe,
autoradiograph

Glass slide
with fixed
cells

Spots show the
chromosomal
location of
the cloned gene

Cells are treated with a fixative, attached to a glass slide and then incubated with ribonuclease and sodium hydroxide to degrade RNA and denature the DNA molecules. Base-pairing between the individual polynucleotide strands is broken down, and the chromosomes unpack to a certain extent, exposing segments of DNA normally enclosed within their structure (Figure 9.5(b)). A sample of the cloned gene is then labelled and applied to the chromosome preparation. Hybridization occurs between the cloned gene and its chromosomal copy, resulting in a dark spot on an autoradiograph. The position of this spot indicates the location of the cloned gene on its chromosome. Although a difficult technique, *in situ* hybridization with radioactively-labelled probes has been used to position a number of genes on the human cytogenetic map.

As an alternative to radioactive labelling, a fluorescent marker can be attached to the probe and hybridization observed directly, using a special type of light microscope. If different fluorochromes are used then two or more genes can be probed at the same time, the different hybridization signals being distinguished by the distinctive colours of their fluorescences. This technique (**FISH – fluorescence *in situ* hybridization**) is also frequently used with probes whose normal chromosomal locations are already known, this being particularly useful for studying cells in which chromosomal rearrangements have occurred. Rearrangements such as chromosome duplications, or the translocation of a segment of one chromosome to another, can be typed relatively quickly by FISH, more quickly than by conventional staining techniques.

(c) Walking along a chromosome from one gene to another
So far we have looked at ways of finding the position on a chromosome of a gene that has already been cloned. Quite frequently the reverse problem arises: the position of the gene is known, but no probe is available for it, so the relevant clone cannot be isolated from a genomic library. **Chromosome walking** allows the desired clone to be identified, but only so long as a clone is already available for a second gene, one that is known to have a chromosomal location near to the desired gene.

Two clone libraries are needed, each prepared with a different restriction endonuclease, say *Eco*RI for one and *Bam*HI for the other. The fragments carried by the clones in the two libraries will overlap (Figure 9.6(a)). To begin the chromosome walk the clone containing the known gene is taken from library A and used to probe library B (Figure 9.6(b)). One or more clones from library B will give positive hybridization signals, showing that the fragments carried by these clones overlap with the fragment carried by the probe. One of these clones from library B is now used to probe library A (Figure 9.6(c)). The original clone, and possibly some others, will hybridize. The cycle is repeated several times, gradually building up a partial restriction map of the overlapping fragments, until the gene being sought is reached (Figure 9.6(d)).

Chromosome walking is a difficult and tedious technique. To date the longest walks that have been achieved are only 200 to 250 kb in length. However, the procedure has had a number of outstanding successes, notably in locating genes responsible for

Figure 9.6 Chromosome walking.

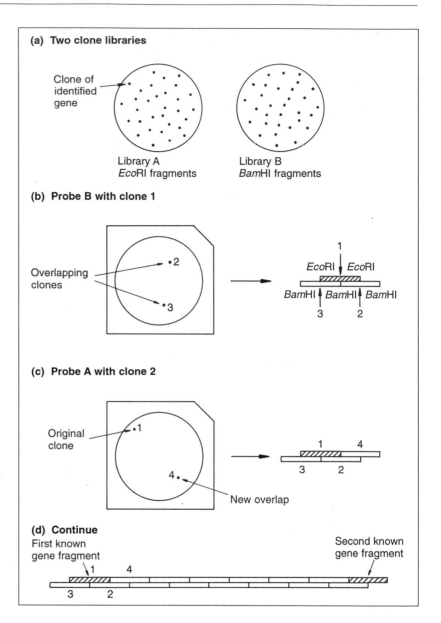

(a) Two clone libraries

Clone of
identified
gene

Library A
*Eco*RI fragments

Library B
*Bam*HI fragments

(b) Probe B with clone 1

Overlapping
clones

•2

•3

1

*Eco*RI   *Eco*RI

*Bam*HI   *Bam*HI   *Bam*HI

3   2

(c) Probe A with clone 2

Original
clone

•1

4•

New overlap

1   4

3   2

(d) Continue

First known
gene fragment

Second known
gene fragment

1   4

3   2

human diseases such as cystic fibrosis, and future refinements will undoubtedly increase the general applicability of the procedure.

9.2 DNA SEQUENCING – WORKING OUT THE STRUCTURE OF A GENE

Probably the most important technique available to the molecular biologist is DNA sequencing, by which the precise order of nucleotides in a piece of DNA can be determined. DNA sequencing methods have been around for 30 years, but only since the late 1970s has rapid and efficient sequencing been possible. Two different techniques were developed almost simultaneously – the chain termination method by F. Sanger and A. R. Coulson in the UK, and the chemical degradation method by A. Maxam and W. Gilbert in the USA. The two techniques are radically different but equally valuable. Both allow DNA sequences of several kb in length to be determined in the minimum of time. The DNA sequence is now the first and most basic type of information to be obtained about a cloned gene.

9.2.1 The Sanger–Coulson method – chain-terminating nucleotides

The chain termination method requires single-stranded DNA and so the molecule to be sequenced is usually cloned into an M13 vector. This is because chain termination sequencing involves the enzymatic synthesis of a second strand of DNA, complementary to an existing template.

(a) **The primer**   The first step in a chain termination sequencing experiment is to anneal a short oligonucleotide primer on to the recombinant M13 molecule (Figure 9.7(a)). This primer will act as the starting point for the complementary strand synthesis reaction carried out by the Klenow fragment of DNA polymerase I, or a related enzyme such as 'Sequenase', a modified version of the DNA polymerase encoded by bacteriophage T7. Remember that these enzymes need a double-stranded region from which to begin strand synthesis (p. 57). The primer anneals to the vector at a position adjacent to the polylinker.

(b) **Synthesis of the complementary strand**   The strand synthesis reaction is started by adding the enzyme plus each of the four deoxynucleotides (dATP, dTTP, dGTP, dCTP). In addition a single modified nucleotide is also included in the reaction mixture. This is a **dideoxynucleotide** (e.g. dideoxyATP) which can be incorporated into the growing polynucleotide strand just as efficiently as the normal nucleotide, but which blocks further strand synthesis. This is because the dideoxynucleotide lacks the hydroxyl group at the 3' position of the sugar component (Figure 9.7(b)). This group is needed for the next nucleotide to be attached; chain termination therefore occurs whenever a dideoxynucleotide is incorporated by the enzyme.

If dideoxyATP is added to the reaction mix, then termination occurs at positions opposite thymidines in the template (Figure

Figure 9.7 Chain termination DNA sequencing.

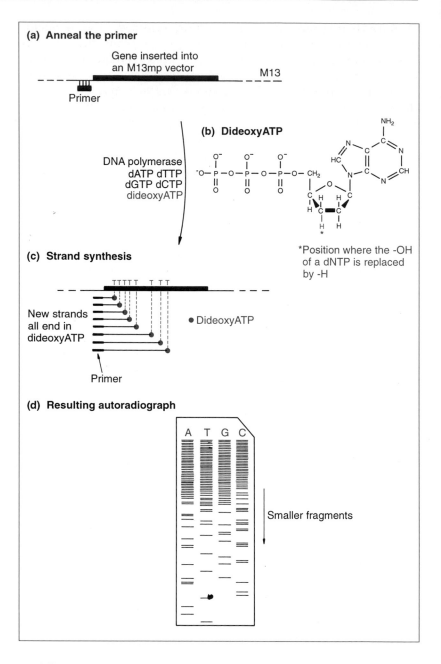

(a) Anneal the primer

Gene inserted into
an M13mp vector   M13

Primer

(b) DideoxyATP

DNA polymerase
dATP dTTP
dGTP dCTP
dideoxyATP

*Position where the -OH
of a dNTP is replaced
by -H

(c) Strand synthesis

New strands
all end in
dideoxyATP

• DideoxyATP

Primer

(d) Resulting autoradiograph

A T G C

Smaller fragments

9.7(c)). But termination does not always occur at the first T as normal dATP is also present and may be incorporated instead of the dideoxynucleotide. The ratio of dATP to dideoxyATP is such that an individual strand may be polymerized for a considerable distance before a dideoxyATP molecule is added. The result is that a family of new strands is obtained, all of different lengths, but each ending in dideoxyATP.

(c) **Four separate reactions result in four families of terminated strands** The strand synthesis reaction is carried out four times in parallel. As well as the reaction with dideoxyATP, there will be one with dideoxyTTP, one with dideoxyGTP, and one with dideoxyCTP. The result is four distinct families of newly synthesized polynucleotides, one family containing strands all ending in dideoxyATP, one of strands ending in dideoxyTTP, etc.

The next step is to separate the components of each family so the lengths of each strand can be determined. This can be achieved by gel electrophoresis, although the conditions have to be carefully controlled as it is necessary to separate strands that differ in length by just one nucleotide. In practice, the electrophoresis is carried out in very thin polyacrylamide gels (less than 0.5 mm thick). The gels contain urea, which denatures the DNA so the newly synthesized strands dissociate from the templates. In addition, the electrophoresis is carried out at a high voltage, so the gel heats up to 60°C and above, making sure the strands do not reassociate in any way.

Each band in the gel contains only a small amount of DNA, so autoradiography has to be used to visualize the results (Figure 9.7(d)). The label is introduced into the new strands by including a radioactive deoxynucleotide (e.g. ^{32}P- or ^{35}S-dATP) in the reaction mixture for the strand synthesis step earlier in the experiment.

(d) **Reading the DNA sequence from the autoradiograph** Reading the sequence is very easy (Figure 9.8). First the band that has moved the furthest is located. This represents the smallest piece of DNA, the strand terminated by incorporation of the dideoxynucleotide at the first position in the template. The track in which this band occurs is noted. Let us say it is track A; the first nucleotide in the sequence is therefore A.

The next most mobile band corresponds to a DNA molecule one nucleotide longer than the first. The track is noted, T in the example shown in Figure 9.8; the second nucleotide is therefore T and the sequence so far is AT.

The process is continued along the autoradiograph until the individual bands become so bunched up that they cannot be separated from one another. Generally it is possible to read a sequence of about 400 nucleotides from one autoradiograph.

9.2.2 The Maxam–Gilbert method – chemical degradation of DNA

There are only a few similarities between the Sanger–Coulson and Maxam–Gilbert methods of DNA sequencing. The Maxam–Gilbert method requires double-stranded DNA fragments, so cloning into an M13 vector is not an essential first step. Neither is a primer

Figure 9.8 Interpreting the autoradiograph produced by a chain termination sequencing experiment. Each track contains the fragments produced by strand synthesis in the presence of one of the four dideoxyNTPs. The sequence is read by identifying the track that each fragment lies in, starting with the one that has moved the furthest, and gradually progressing up through the autoradiograph.

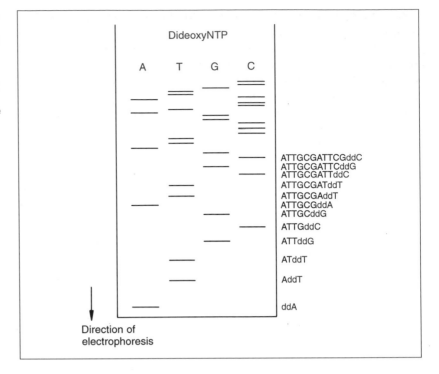

needed, because the basis of the Maxam–Gilbert technique is not synthesis of a new strand, but cleavage of the existing DNA molecule using chemical reagents that act specifically at a particular nucleotide.

There are several variations of the Maxam–Gilbert method, differing in details such as the way in which the labelled DNA is obtained, and the exact nature of the cleavage reagents that are used. Most of these reagents are very toxic – chemicals that cleave DNA molecules in the test-tube will do the same in the body, and great care must be taken when using them.

(a) Performing a Maxam–Gilbert sequencing experiment The following is a popular version of the Maxam–Gilbert technique. The double-stranded DNA fragment to be sequenced is first labelled by attaching a radioactive phosphorus group to the 5′ end of each strand (Figure 9.9(a)). Dimethyl sulphoxide is then added and the labelled DNA sample heated to 90°C. This results in breakdown of the base-pairing and dissociation of the DNA molecule into its two component strands. The two strands are separated from one another by gel electrophoresis (Figure 9.9(b)), which works on the basis that one of the strands probably contains more purine nucleotides than the other and will therefore be slightly heavier. One strand is purified from the gel and divided into four samples, each of which is treated with one of the cleavage reagents.

In fact, the first set of reagents to be added cause a chemical modification in the nucleotides they are specific for, making the strand susceptible to cleavage at that nucleotide when an additional chemical, piperidine, is added (Figure 9.9(c)). The modification and cleavage reactions are carried out under conditions that result in only one breakage per strand.

Some of the cleaved fragments retain the ^{32}P label at their 5′ ends. After electrophoresis, using the same special conditions as for chain termination sequencing, the bands visualized by autoradiography will represent these labelled fragments. The nucleotide

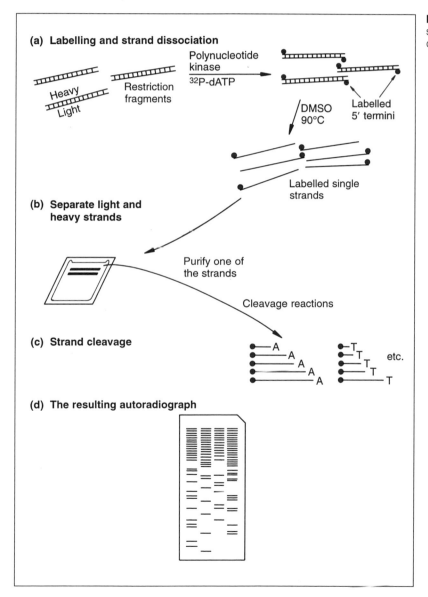

Figure 9.9 One version of DNA sequencing by the chemical degradation method.

(a) **Labelling and strand dissociation**

Heavy / Light

Restriction fragments

Polynucleotide kinase
^{32}P-dATP

Labelled 5′ termini

DMSO 90°C

Labelled single strands

(b) **Separate light and heavy strands**

Purify one of the strands

Cleavage reactions

(c) **Strand cleavage**

etc.

(d) **The resulting autoradiograph**

sequence can now be read from the autoradiograph exactly as for a chain termination experiment (Figure 9.9(d)).

9.2.3 Building up a long DNA sequence

A single DNA sequencing experiment, whether using the Sanger–Coulson or the Maxam–Gilbert method, produces only about 400 nucleotides of sequence. But most genes are much longer than this; how can a sequence of several kb be obtained?

The answer is to perform DNA sequencing experiments with a lot of different fragments, all derived from a single larger DNA molecule (Figure 9.10). These fragments should overlap, so the

Figure 9.10 Building up a long DNA sequence from a series of short overlapping ones.

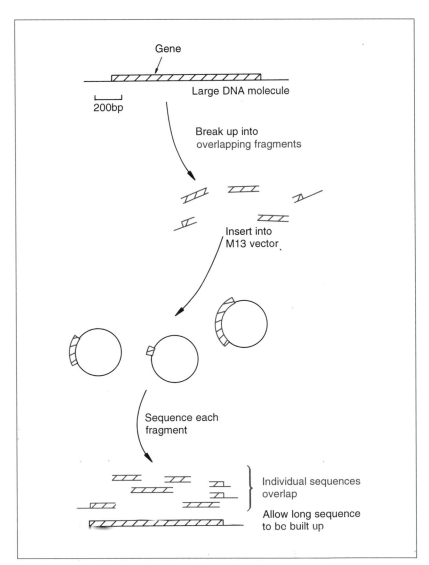

individual DNA sequences will themselves overlap. The overlaps can be located, either by eye or using a computer, and the master sequence gradually built up.

There are several ways of producing the overlapping fragments. For instance, the DNA molecule could be cleaved with two different restriction endonucleases, producing one set of fragments with say *Sau*3A and another with *Alu*I (Figure 9.11). This is the traditional method of producing overlapping sequences but suffers from the drawback that the restriction sites may be inconveniently placed and individual fragments may be too long to be completely sequenced. Often four or five different restriction endonucleases will have to be used to clear up all the gaps in the master sequence. However, this is still the easiest and most foolproof method of obtaining a fully contiguous sequence.

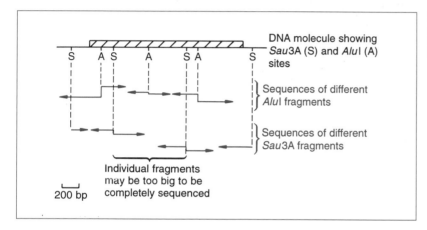

Figure 9.11 Building up a long DNA sequence by determining the sequences of overlapping restriction fragments.

9.2.4 The achievements of DNA sequencing

The first DNA molecule to be completely sequenced was the 5386 nucleotide genome of bacteriophage φX174, which was completed in 1975. This was quickly followed by sequences for SV40 virus (5243 bp) in 1977 and pBR322 (4363 bp) in 1978. Gradually sequencing was applied to larger molecules. Professor Sanger's group published the sequence of the human mitochondrial genome (16.6 kb) in 1981 and of bacteriophage λ (49 kb) in 1982. Nowadays sequences of 10–20 kb are routine and most research laboratories have the necessary expertise to generate this amount of information. The pioneering projects today are the massive genome initiatives, each aimed at obtaining the nucleotide sequence of the entire genome of a particular organism. The Human Genome Project is making rapid progress and complete chromosome sequences are already available for *Saccharomyces cerevisiae*. The challenges of compiling, interpreting and understanding the wealth of sequence data provided by these projects will undoubtedly be the major preoccupation of molecular biologists in the early decades of the 21st century.

9.3 USING CLONED GENES TO STUDY GENOME STRUCTURE

Is it always necessary to obtain a nucleotide sequence in order to study the structure of a gene or a genome? The answer is no, as less detailed, but nonetheless valuable information can also be obtained by a variety of indirect methods, these techniques having the advantage of being relatively rapid and enabling comparisons to be made between multiple samples. We will consider two of these techniques, **restriction fragment length polymorphism (RFLP)** analysis and **genetic fingerprinting**.

9.3.1 Restriction fragment length polymorphism analysis

If two DNA molecules are very similar, but have a few small differences in their nucleotide sequences, then the fact that they are not identical may be apparent from a comparison of their restriction maps. Three possibilities (not the only ones) are illustrated in Figure 9.12, each of which results in a restriction fragment length polymorphism that would be detected if the restriction fragments concerned could be visualized by agarose gel electrophoresis.

The problem is that with most organisms the individual fragments can not be seen after electrophoresis. Restriction of human DNA with *Eco*RI, for instance, results in approximately 700 000 fragments, producing a smear in an agarose gel from which it is impossible to distinguish just one altered band.

In practice, a strategy based on Southern hybridization is therefore used. The restriction digest is electrophoresed in the normal way but, rather than being visualized by ethidium bromide staining, the smear of fragments is transferred to a nitrocellulose or nylon membrane by Southern blotting (Figure 9.13). Hybridization analysis is then carried out, using as the probe a clone that spans the region of interest. The probe hybridizes to the relevant region, 'lighting up' the appropriate restriction fragments on the resulting autoradiograph. If an RFLP is present then it will be clearly visible when the banding pattern is compared with that obtained with the unmutated gene.

(a) **Applications of RFLP analysis** RFLP analysis has a number of applications in molecular biology, the most important being its use in screening programmes for human gene mutations that might lead to genetic disease. This analysis is possible if the mutation responsible for the genetic disease also causes an RFLP, as will

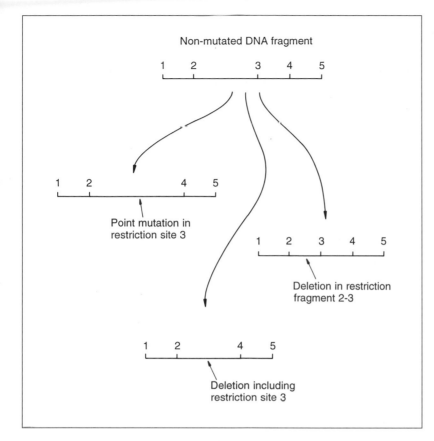

Figure 9.12 Three mutations and the effects they will have on the restriction map of a DNA molecule.

occur with all deletion and insertion mutations, as well as those point mutations that result in loss of a restriction site. The presence of the RFLP is then a direct indication that the gene is defective and that the individual may suffer from the resulting disease. Unfortunately, only a few genetic diseases can be diagnosed in this way, not because genetic defects that result in RFLPs are rare, but because as yet we do not know enough about the structures of the relevant genes to know what RFLPs to look for. Despite this, the method has proved useful for diagnosing various types of thalassaemia, which result from defects in the globin genes, and also in diagnosis of some types of haemophilia.

An alternative and more generally useful method of genetic screening is **RFLP linkage analysis**. This does not depend on the existence of an RFLP that is a direct consequence of the genetic defect, but requires merely that a recognizable RFLP is present somewhere in the vicinity of the defective gene. If this RFLP is close enough then it will be inherited with the defective gene, as it is very unlikely that a recombination event during meiosis will separate the RFLP from the gene. The presence of the RFLP will therefore be indicative of the presence of the defective gene.

Figure 9.13 Examining a sample of human DNA for a restriction fragment length polymorphism.

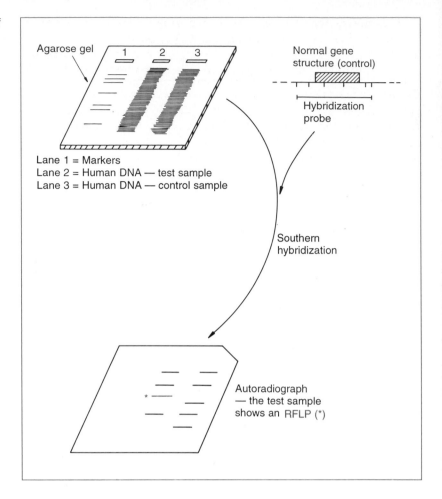

9.3.2 Genetic fingerprinting

Genetic fingerprinting is a specialized form of RFLP analysis that enables a more general picture of genome structure to be obtained. Rather than concentrating on just a single RFLP, genetic fingerprinting examines a large number of RFLPs at once. One way of doing this would be to use a mixed probe, made up of several different DNA molecules targeting a number of different parts of the genome, but most forms of genetic fingerprinting are more subtle than this. The targeted regions are tandem repeats, each of which consists of a short sequence, up to about 64 bp in length, that is repeated numerous times in succession (Figure 9.14(a)). These tandem repeats have two features that form the basis to genetic fingerprinting:

1. Arrays of repeats with similar or identical core sequences are located at a number of places in the genome (Figure 9.14(b)).

This means that a single probe, recognizing the core sequence, can be used to type a number of RFLPs at once.

2. The number of repeat units in a single array is hypervariable. This means that the arrays display RFLPs at a high frequency.

Genetic fingerprinting is carried out in exactly the same way as standard RFLP analysis. Genomic DNA is restricted, electrophoresed, transferred to a nitrocellulose or nylon membrane, and hybridized to the probe. The difference is that the resulting autoradiograph does not show a simple pattern of four or five bands, as drawn in Figure 9.13. Instead the autoradiograph dis-

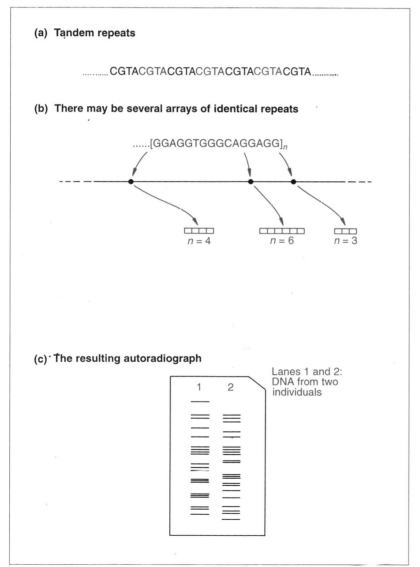

(a) Tandem repeats

............CGTACGTACGTACGTACGTACGTACGTA............

(b) There may be several arrays of identical repeats

......[GGAGGTGGGCAGGAGG]$_n$

$n = 4$ $n = 6$ $n = 3$

(c) The resulting autoradiograph

1   2

Lanes 1 and 2: DNA from two individuals

Figure 9.14 Genetic fingerprinting.

plays a complex pattern of bands (Figure 9.14(c)), reflecting the fact that the probe hybridizes to arrays from numerous loci within the genome. The banding patterns obtained from two different individuals are unlikely to be the same (unless they are monozygotic twins) because of the hypervariability of each repeat array.

(a) Applications of genetic fingerprinting The most celebrated application of genetic fingerprinting is in forensic science. If the appropriate probes are used then the chances of two individuals who are not monozygotic twins having the same banding pattern can be of the order of one in several million. This means that a match between the genetic fingerprint of an alleged wrongdoer and that obtained from blood or hair recovered from the scene of the crime is considered highly suspicious. Genetic fingerprinting is also used in other courtroom dramas concerning paternity ormaternity disputes. Although an individual's genetic fingerprint is unique the banding pattern shows similarities to both the mother and father, from whom the DNA sequences were inherited (Figure 9.15). Comparisons of genetic fingerprints can therefore determine if two individuals are related. This is important not only with human fingerprints but also with animals such as valuable racehorses and pedigree dogs, and has found scientific applications in studies of bird and animal populations.

Figure 9.15 Parents and offspring have similar genetic fingerprints. Each band in the daughter's genetic fingerprint is inherited from either the father or the mother.

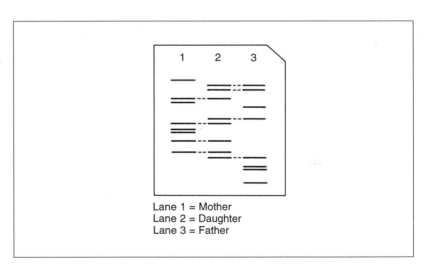

Lane 1 = Mother
Lane 2 = Daughter
Lane 3 = Father

FURTHER READING

Southern, E. M. (1975) Detection of specific sequences among DNA fragments separated by gel electrophoresis. *Journal of Molecular Biology*, **98**, 503–7 – Southern hybridization.

Carle, G. E. and Olson, M. V. (1985) An electrophoretic karyotype for yeast. *Proceedings of the National Academy of Sciences, USA*, **82**, 3756–60 – an application of OFAGE.

Bender, W. *et al.* (1983) Chromosome walking and jumping to isolate DNA from the *Ace* and *rosy* loci and the bithorax complex in *Drosophila melanogaster*. *Journal of Molecular Biology*, **168**, 17–33 – one of the first examples of chromosome walking.

Sanger, F. *et al.* (1977) DNA sequencing with chain-terminating inhibitors. *Proceedings of the National Academy of Sciences, USA*, **74**, 5463–7.

Maxam, A. and Gilbert, W. (1977) A new method of sequencing DNA. *Proceedings of the National Academy of Sciences, USA*, **74**, 560–4.

Brown, T. A. (1994) *DNA Sequencing: The Basics*. Oxford University Press, Oxford – everything you need to know about DNA sequencing.

Johnston, M. *et al.* (1994) Complete nucleotide sequence of *Saccharomyces cerevisiae* chromosome VIII. *Science*, **265**, 2077–82.

Oliver, S. G. (1994) Back to bases in biology. *Nature*, **368**, 14–15 – a discussion of the status and potential importance of genome sequencing projects.

Jeffreys, A. J., Wilson, V. and Thein, S. L. (1985) Individual-specific 'fingerprints' of human DNA. *Nature*, **316**, 76–9.

10 Studying gene expression

All genes have to be expressed in order to function. The first step in expression is transcription of the gene into a complementary RNA strand (Figure 10.1(a)). For some genes – for example those coding for transfer RNA and ribosomal RNA molecules – the transcript itself is the functionally important molecule. For other genes the transcript is translated into a protein molecule.

To understand how a gene is expressed, the RNA transcript must be studied. In particular, the molecular biologist will want to know whether the transcript is a faithful copy of the gene, or whether segments of the gene are missing from the transcript (Figure 10.1(b)). These missing pieces are called introns and considerable interest centres on their structure and possible function. In addition to introns, the exact locations of the start and end points of transcription are important. Most transcripts are copies not only of the gene itself, but also of the nucleotide regions either side of it (Figure 10.1(c)). The signals that determine the start and finish of the transcription process are only partly understood, but their positions must be located if the expression of a cloned gene is to be studied.

In this chapter we will begin by looking at the methods used for **transcript analysis**. These methods can be used to determine if a cloned gene contains introns and to map the positions of the start and end points for transcription. Then we will briefly consider a few of the numerous techniques developed in recent years for examining how expression of a cloned gene is regulated. These techniques are important as aberrations in gene regulation underlie many clinical disorders. Finally, we will tackle the difficult problem of how to identify the translation product of a cloned gene.

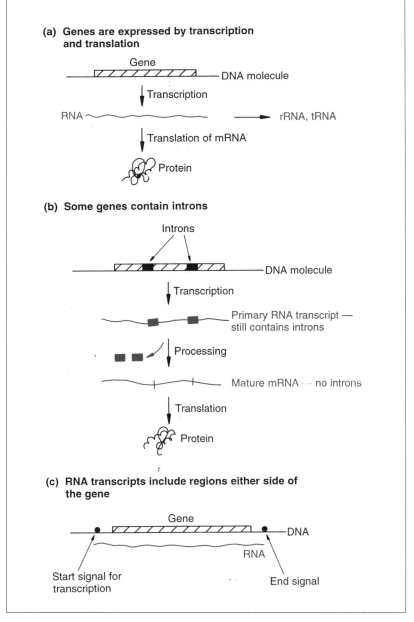

Figure 10.1 Some fundamentals of gene expression.

10.1 STUDYING THE TRANSCRIPT OF A CLONED GENE

Most methods of transcript analysis are based on hybridization between the RNA transcript and a fragment of DNA containing the relevant gene. Nucleic acid hybridization occurs just as readily between complementary DNA and RNA strands as between single-stranded DNA molecules. The resulting DNA–RNA hybrid can be analysed by electron microscopy or with single-strand specific nucleases.

10.1.1 Electron microscopy of nucleic acid molecules

Electron microscopy can be used to visualize nucleic acid molecules, so long as the polynucleotides are first treated with chemicals that increase their apparent diameter. Untreated molecules are simply too thin to be seen.

Usually the DNA molecules are mixed with a protein such as cytochrome c which binds to the polynucleotides, coating the strands in a thick shell. The coated molecules must be stained with uranyl acetate or some other electron-dense material to enhance the appearance of the preparation (Figure 10.2). Quite spectacular views of nucleic acid molecules can be obtained.

In the past, electron microscopy has been used primarily to analyse hybridization between different DNA molecules, but in recent years the technique has become increasingly important in the study of DNA–RNA hybrids. It is particularly useful for determining if a gene contains introns. Consider the appearance of a hybrid between a DNA strand, containing a gene, and its RNA transcript. If the gene contains introns then these regions of the DNA strand will have no homology with the RNA transcript and

Figure 10.2 Preparing a DNA molecule for observation with the electron microscope.

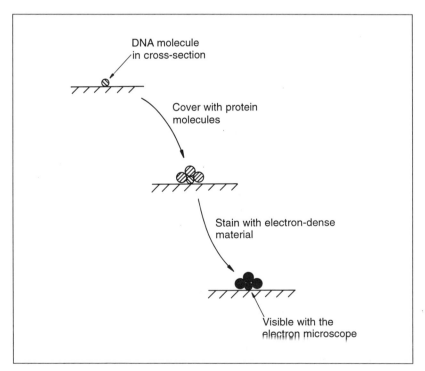

so can not base pair. Instead they 'loop out', giving a characteristic appearance when observed with the electron microscope (Figure 10.3). The number and positions of these loops correspond directly to the number and positions of the introns in the gene. Further information can then be obtained by sequencing the gene and looking for the characteristic features that mark the boundaries of introns. If a cDNA clone is available then its sequence, which of course lacks the introns, can be compared with the gene sequence to locate the introns with precision.

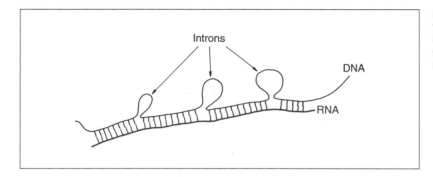

Figure 10.3 The appearance under the electron microscope of a DNA–RNA hybrid formed between a gene containing an intron and its processed transcript.

10.1.2 Analysis of DNA–RNA hybrids by nuclease treatment

The second method for studying a DNA–RNA hybrid involves a single-strand specific nuclease such as S1 (p. 54). This enzyme degrades single-stranded DNA or RNA polynucleotides, including single-stranded regions at the end of predominantly double-stranded molecules, but has no effect on double-stranded DNA or on DNA–RNA hybrids. If a DNA molecule containing a gene is hybridized to its RNA transcript, and then treated with S1 nuclease, the non-hybridized single-stranded DNA regions at each end of the hybrid will be digested, along with any looped-out introns (Figure 10.4). The result is a completely double-stranded hybrid. The single-stranded DNA fragments protected from S1 nuclease digestion can be recovered if the RNA strand is degraded by treatment with alkali.

Unfortunately the manipulations shown in Figure 10.4 are not very informative. The sizes of the protected DNA fragments could be measured by gel electrophoresis, but this does not allow their order or relative positions to be determined. However, a few subtle modifications to the technique allows the precise start and end points of the transcript and of any introns it contains to be mapped on to the DNA sequence.

In the example shown in Figure 10.5, a *Sau*3A fragment that contains 100 bp of coding region, along with 300 bp of the leader

Figure 10.4 The effect of S1 nucle-ase on a DNA–RNA hybrid.

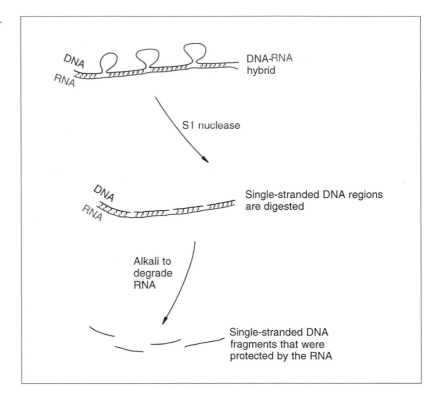

sequence preceding the gene, has been cloned into an M13 vector and obtained as a single-stranded molecule. A sample of the RNA transcript is added and allowed to anneal to the DNA molecule. The DNA molecule will still primarily be single-stranded but will now have a small region protected by the RNA transcript. All but this protected region is digested by S1 nuclease and the RNA degraded with alkali, leaving a short single-stranded DNA fragment. If these manipulations are examined closely it will become clear that the size of this single-stranded fragment corresponds to the distance between the transcription start point and the right-hand *Sau*3A site. The size of the single-stranded fragment is there-fore determined by gel electrophoresis and this information is used to locate the transcription start point on the DNA sequence. Exactly the same strategy could locate the end point of transcription and the junction points between introns and exons: the only difference would be the position of the restriction site chosen to delimit one end of the protected single-stranded DNA fragment.

10.1.3 Other techniques for studying RNA transcripts

Electron microscopy and S1 nuclease analysis are not the only methods available for studying RNA transcripts. In recent years a broad range of RNA manipulative techniques have been developed, including reliable methods for direct sequencing of RNA molecules. These methods enable RNA transcription to be studied in ways that have been impossible in the past and, in particular, have resulted in a better understanding of how transcripts are processed. The primary transcript that is copied directly from the gene may undergo a complex series of modification events before the mature mRNA molecule is produced. As well as removal of introns a transcript may be edited by insertion of new nucleotides and by alteration of existing ones. Some RNA molecules may even

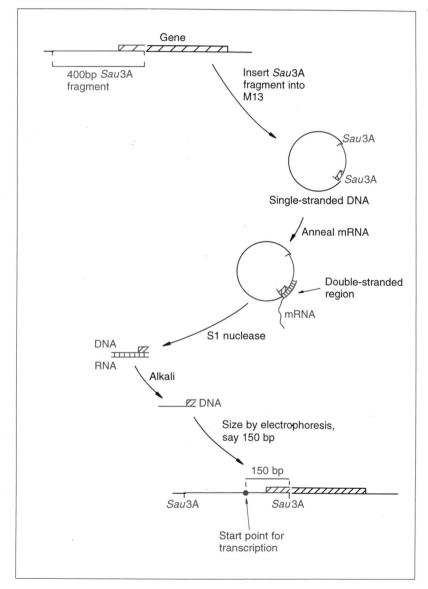

Figure 10.5 Locating a transcription start point by S1 nuclease mapping.

possess catalytic activity. These exciting discoveries are stimulating the development of new methods for transcript analysis and this is one of many areas of recombinant DNA technology that is progressing rapidly.

10.2 STUDYING THE REGULATION OF GENE EXPRESSION

Few genes are expressed all the time – most are subject to regulation and are switched on only when their gene product is required by the cell. The simplest gene regulation systems are found in bacteria, such as *E. coli*, which can regulate expression of genes for biosynthetic and metabolic processes, so that gene products that are not needed are not synthesized. For instance the genes coding for the enzymes involved in tryptophan biosynthesis can be switched off when there are abundant amounts of tryptophan in the cell, and switched on again when tryptophan levels drop. Similarly, genes for the utilization of sugars such as lactose are activated only when the relevant sugar is there to be metabolized. In higher organisms gene regulation is more complex because there are many more genes that are subject to control. Differentiation of cells in tissues and organs involves wholesale changes in gene expression patterns, with many of the changes being irreversible. The process of development from fertilized egg cell to adult also involves gene regulation and requires coordination between different cells as well as time-dependent changes in gene expression.

Many of the problems in gene regulation require a classical genetic approach; genetics enables genes that control regulation to be identified, can define the biochemical signals that influence gene expression, and explores the interactions between different genes and gene families. It is for this reason that most of the breakthroughs in understanding development in higher organisms have started with studies of the fruit-fly *Drosophila melanogaster*. Gene cloning complements classical genetics as it provides much more detailed information on the molecular events involved in regulating the expression of a single gene. We now know that a gene subject to regulation has one or more control sequences in its upstream region (Figure 10.6) and that the gene is switched on and off by the attachment of regulatory proteins to these sequences. A regulatory protein may repress gene expression, in which case the gene is switched off when the protein is bound to the control sequence, or alternatively the protein may have a positive or enhancing role, switching on or increasing expression of the target gene. In this section we will examine methods for locating control sequences and determining their roles in regulating gene expression. The more difficult questions, such as what controls the binding of the

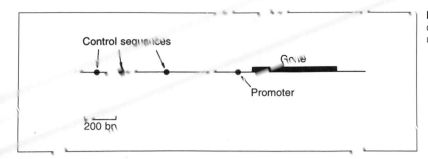

Control sequences

Gene

Promoter

200 bp

regulatory protein to the control sequence, is a problem for genetics and beyond the scope of the simple gene cloner.

10.2.1 Identifying protein-binding sites on a DNA molecule

A control sequence is a region of DNA that can bind a regulatory protein. It should therefore be possible to identify control sequences upstream of a cloned gene by searching the relevant region for protein-binding sites. There are two different approaches. In the first, a DNA fragment that has a protein bound to it is identified by virtue of its increased molecular mass, and in the second the region to which the protein is bound is delineated by its resistance to endonuclease cleavage. In each case the first step is to form a complex between the regulatory protein and the DNA molecule under study. It is very unlikely that the regulatory protein will be available in pure form, so usually the starting material is an unfractionated extract of nuclear protein (remember that gene regulation occurs in the nucleus).

(a) **Gel retardation of DNA–protein complexes** Proteins are quite substantial structures and a protein attached to a DNA molecule results in a large increase in overall molecular mass. If this increase can be detected then a DNA fragment containing a protein-binding site will have been identified. In practice a DNA fragment carrying a bound protein is identified by gel electrophoresis, as it has a lower mobility than the uncomplexed DNA molecule (Figure 10.7). The procedure is referred to as **gel retardation**.

For gel retardation to be of value the region containing the control sequence must be digested with a restriction endonuclease before mixing with the regulatory protein (Figure 10.8). The location of the control sequence is then determined by finding the position on the restriction map of the fragment that is retarded during gel electrophoresis. The precision with which the control sequence can be located depends on how detailed the restriction map is and how conveniently placed the restriction sites are. A single control sequence may be less than 10 bp in size, so rarely will gel

Figure 10.7 A bound protein will lower the mobility of a DNA fragment during gel electrophoresis.

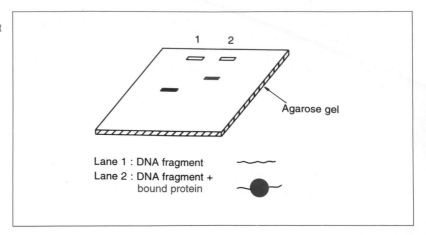

·retardation be able to pinpoint it exactly. A more precise technique is therefore needed to delineate the exact position of the protein-binding sequence within the fragment identified by gel retardation.

(b) Footprinting with DNase I The procedure generally called **footprinting** is the desired supplement to gel retardation. Foot-

Figure 10.8 Carrying out a gel retardation experiment.

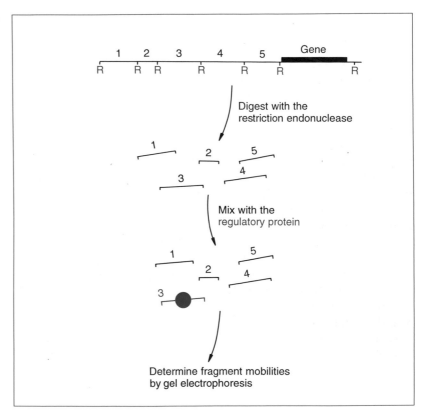

printing works on the basis that the interaction with a regulatory protein protects the DNA in the region of a control sequence from the degradative action of an endonuclease such as DNase I (Figure 10.9). This phenomenon can be used to locate with some precision the protein-binding site on the DNA molecule.

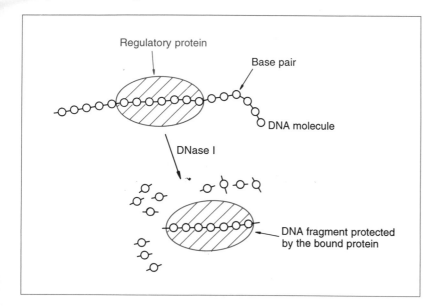

Figure 10.9 A bound protein protects a region of a DNA molecule from degradation by a nuclease such as DNase I.

The DNA fragment being studied is first labelled at one end with a radioactive marker, and then complexed with the regulatory protein (Figure 10.10(a)). DNase I is added, but the amount used is limited so that complete degradation of the DNA fragment does not occur. Instead the aim is to cut each molecule at just a single phosphodiester bond (Figure 10.10(b)). If the DNA fragment has no protein attached to it then the result of this treatment will be a family of labelled fragments, differing in size by just one nucleotide each. After separation on a polyacrylamide gel the family of fragments appears as a ladder of bands on an autoradiograph (Figure 10.10(c)). However, the bound protein protects certain phosphodiester bonds from being cut by DNase I, meaning that in this case the family of fragments is not complete, as the fragments resulting from cleavage within the control sequence are absent. Their absence shows up on the autoradiograph as a 'footprint', clearly seen in Figure 10.10(c). The position of the control sequence within the DNA molecule can now be worked out from the sizes of the fragments on either side of the footprint.

Figure 10.10 DNase I footprinting.

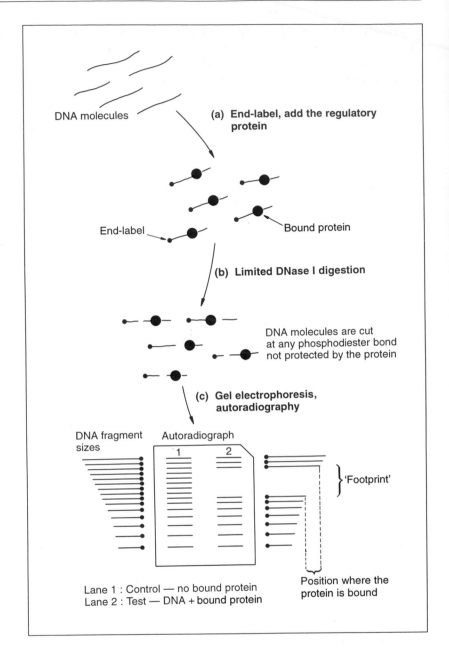

DNA molecules

(a) End-label, add the regulatory protein

End-label

Bound protein

(b) Limited DNase I digestion

DNA molecules are cut at any phosphodiester bond not protected by the protein

(c) Gel electrophoresis, autoradiography

DNA fragment sizes

Autoradiograph

1   2

'Footprint'

Lane 1 : Control — no bound protein
Lane 2 : Test — DNA + bound protein

Position where the protein is bound

10.2.2 Identifying control sequences by deletion analysis

Gel retardation and footprinting experiments can locate possible control sequences upstream of a cloned gene but provide no information on the function of the individual sequences. **Deletion analysis** is a totally different approach that can not only locate control sequences (though only with the precision of gel retardation), but importantly can also indicate the function of each sequence.

The technique depends on the assumption that deletion of the control sequence will result in a change in the way in which expression of the cloned gene is regulated (Figure 10.11). For instance, deletion of a sequence that represses expression of a gene should result in that gene being expressed at a higher level. Similarly, tissue-specific control sequences can be identified as their deletion results in the target gene being expressed in tissues other than the correct one.

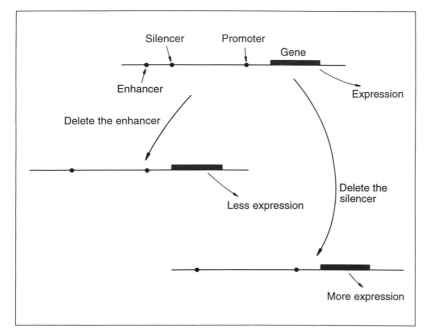

Figure 10.11 The principle behind deletion analysis.

(a) Reporter genes   Before carrying out deletion analysis a way must be found to assay the effect of a deletion on expression of the cloned gene. The effect will probably only be observed when the gene is cloned into the species it was originally obtained from: it will be no good assaying for light-regulation of a plant gene if the gene is cloned in a bacterium.

Cloning vectors have now been developed for most organisms (Chapter 7) so cloning the gene under study back into its host should not cause a problem. The difficulty is that in most cases the host already possesses a copy of the cloned gene. How can changes in the expression pattern of the cloned gene be distinguished from the normal pattern of expression displayed by the host's copy of the gene? The answer is to use a **reporter gene**. This is a test gene that is fused to the upstream region of the gene under study (Figure 10.12), replacing the original cloned gene. When cloned into the host organism the expression pattern of the reporter gene will exactly mimic

Figure 10.12 A reporter gene.

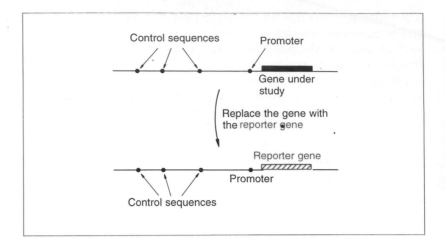

that of the original gene, as the reporter gene is under the influence of exactly the same control sequences as the original gene.

The reporter gene must be chosen with care. The first criterion is that the reporter gene must code for a phenotype not already displayed by the host organism. The phenotype of the reporter gene must be relatively easy to detect after it has been cloned into the host, and ideally it should be possible to assay the phenotype quantitatively. These criteria have not proved difficult to meet and a variety of different reporter genes have been used in studies of gene regulation; a few examples are listed in Table 10.1.

(b) Carrying out a deletion analysis   Once a reporter gene has been chosen and the necessary construction made, carrying out a

Table 10.1   A few examples of reporter genes used in studies of gene regulation in higher organisms

Gene	Gene product	Assay
lacZ	β-Galactosidase	Histochemical test
neo	Neomycin phosphotransferase	Kanamycin resistance
cat	Chloramphenicol acetyltransferase	Chloramphenicol resistance
dhfr	Dihydrofolate reductase	Methotrexate resistance
aphIV	Hygromycin phosphotransferase	Hygromycin resistance
lux	Luciferase	Bioluminescence
uidA	β-Glucuronidase	Histochemical test

All of these genes are obtained from *E. coli*, except for *lux* which has three sources: the luminescent bacteria *Vibrio harveyii* and *V. fischeri*, and the firefly *Photinus pyralis.*

deletion analysis is fairly straightforward. Deletions can be made in the upstream region of the construct by any one of several strategies, a simple example being shown in Figure 10.13. The effect of the deletion is then assessed by cloning the deleted construct into the host organism and determining the pattern and extent of expression of the reporter gene. An increase in expression will imply that a repressing or silencing sequence has been removed, a decrease will indicate removal of an activator or enhancer, and a change in tissue-specificity (as shown in Figure 10.13) will pinpoint a tissue-responsive control sequence.

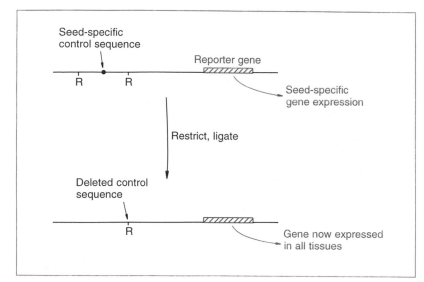

Figure 10.13 Deletion analysis. A reporter gene has been attached to the upstream region of a seed-specific gene from a plant. Removal of the restriction fragment bounded by the sites 'R' deletes the control sequence that mediates seed-specific gene expression, so that the reporter gene is now expressed in all tissues of the plant.

The results of a deletion analysis project have to be interpreted very carefully. Complications may arise if a single deletion removes two closely linked control sequences or, as is fairly common, two distinct control sequences cooperate to produce a single response. Despite these potential difficulties deletion analyses, in combination with studies of protein-binding sites, have provided important information about how the expression of individual genes is regulated, and have supplemented and extended the more broadly based genetic analyses of differentiation and development.

10.3 IDENTIFYING AND STUDYING THE TRANSLATION PRODUCT OF A CLONED GENE

In the last few years gene cloning has become increasingly useful in the study not only of genes themselves but also of the proteins coded by cloned genes. Investigations into protein structure and function have benefited greatly from new techniques that allow mutations to be introduced at specific points in a cloned gene, resulting in directed changes in the structure of the encoded protein.

Before considering these procedures we should first look at the more mundane problem of how to identify the protein coded by a cloned gene. In many cases this analysis is not necessary as the protein will have been characterized long before the gene cloning experiment is performed. On the other hand, there are occasions when the translation product of a cloned gene has not been identified. A method for isolating the protein is then needed.

10.3.1 HRT and HART can identify the translation product of a cloned gene

Two related techniques, **hybrid-release translation (HRT)** and **hybrid-arrest translation (HART)**, are used to identify the translation product encoded by a cloned gene. Both depend on the ability of purified mRNA to direct synthesis of proteins in **cell-free translation systems**. These are cell extracts, usually prepared from germinating wheat seeds or from rabbit reticulocyte cells (both of which are exceptionally active in protein synthesis) and containing ribosomes, tRNAs and all the other molecules needed for protein synthesis. The mRNA sample is added to the cell-free translation system, along with a mixture of the 20 amino acids found in proteins, one of which is labelled (often ^{35}S-methionine is used). The mRNA molecules are translated into a mixture of radioactive proteins (Figure 10.14), which can be separated by gel electrophoresis and visualized by autoradiography. Each band represents a single protein coded by one of the mRNA molecules present in the sample.

HRT and HART work best when a cDNA clone prepared directly from the mRNA sample is available. For HRT the cDNA is denatured, immobilized on a nitrocellulose or nylon membrane, and incubated with the mRNA sample (Figure 10.15). The specific mRNA counterpart of the cDNA hybridizes and remains attached to the membrane. After discarding the unbound molecules, the hybridized mRNA is recovered and translated in a cell-free system. This provides a pure sample of the protein coded by the cDNA.

HART is slightly different in that the denatured cDNA is added directly to the mRNA sample (Figure 10.16). Hybridization again occurs between the cDNA and its mRNA counterpart but in this case the unbound mRNA is not discarded. Instead the entire sample is translated in the cell-free system. The hybridized mRNA is unable to direct translation, so all the proteins except the one coded by the cloned gene are synthesized. The cloned gene's translation product is therefore identified as the protein that is absent from the autoradiograph.

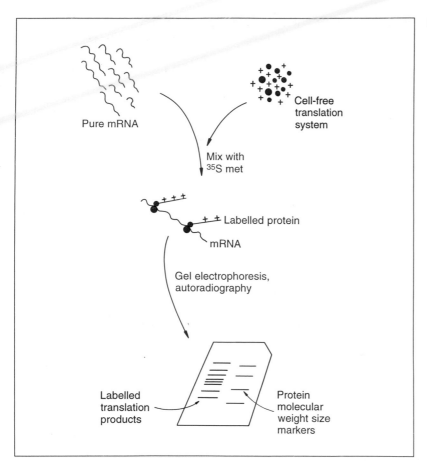

Figure 10.14 Cell-free translation.

10.3.2 Analysis of proteins by *in vitro* mutagenesis

Although HRT and HART can identify the translation product of a cloned gene, these techniques tell us little about the protein itself. The major questions asked by the molecular biologist today centre on the relationship between the structure of a protein and its mode of activity. The best way of tackling these problems is to induce a mutation in the gene coding for the protein and then to determine what effect the change in amino acid sequence has on the properties of the translation product (Figure 10.17). However, under normal circumstances mutations occur randomly and a large number may have to be screened before one that gives useful information is found. A solution to this problem is provided by *in vitro* mutagenesis, a technique that enables a directed mutation to be made at a specific point in a cloned gene.

Figure 10.15 Hybrid-release translation.

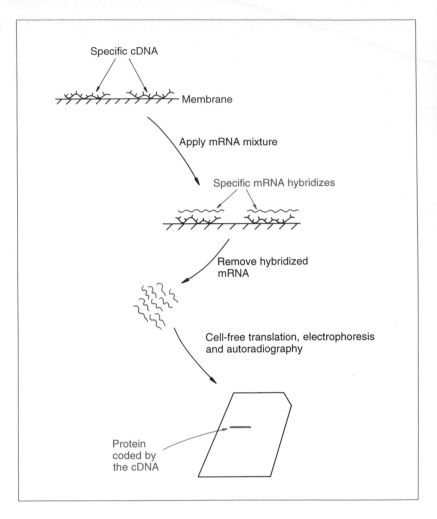

Specific cDNA

Membrane

Apply mRNA mixture

Specific mRNA hybridizes

Remove hybridized mRNA

Cell-free translation, electrophoresis and autoradiography

Protein coded by the cDNA

(a) Different types of *in vitro* mutagenesis techniques An almost unlimited variety of DNA manipulations can be used to introduce mutations into cloned genes. The following are the simplest.

1. A restriction fragment can be deleted (Figure 10.18(a)).
2. The gene can be opened at a unique restriction site, a few nucleotides removed with a double-strand specific endonuclease such as Bal31 (p. 53), and the gene religated (Figure 10.18(b)).
3. A short oligonucleotide can be inserted at a restriction site (Figure 10.18(c)). The sequence of the oligonucleotide can be such that the additional stretch of amino acids inserted into the protein produces, for example, a new structure such as an α-helix, or destabilizes an existing structure.

Although potentially useful, these manipulations depend on the fortuitous occurrence of a restriction site at the area of interest in the cloned gene. **Oligonucleotide-directed mutagenesis** is a more versatile technique that can introduce a mutation at any point in the gene.

(b) Using an oligonucleotide to introduce a point mutation in a cloned gene   For oligonucleotide-directed mutagenesis the gene must usually be obtained in a single-stranded form and so is generally cloned into an M13 vector. The single-stranded DNA is purified and the region to be mutated identified by DNA sequencing. A short oligonucleotide is then synthesized, complementary to the relevant region, but containing the desired nucleotide alteration (Figure 10.19(a)). Despite this mismatch the oligonucleotide will anneal to the single-stranded DNA and act as a primer for comple-

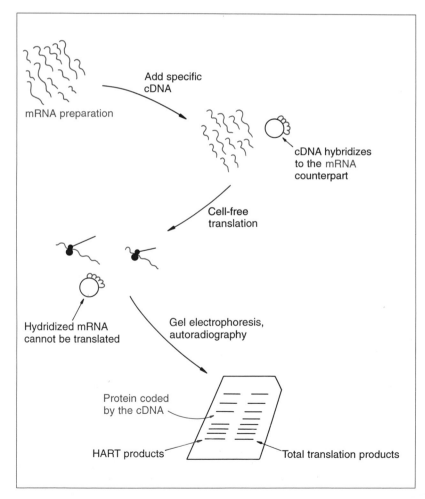

Figure 10.16 Hybrid-arrest translation.

mRNA preparation

Add specific cDNA

cDNA hybridizes to the mRNA counterpart

Cell-free translation

Hydridized mRNA cannot be translated

Gel electrophoresis, autoradiography

Protein coded by the cDNA

HART products

Total translation products

Figure 10.17 A mutation may change the amino acid sequence of a protein, possibly affecting its properties.

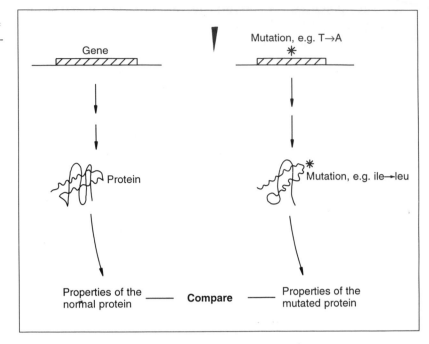

mentary strand synthesis using the Klenow fragment of DNA polymerase I (Figure 10.19(b)). This strand-synthesis reaction is continued until an entire new strand is made and the recombinant molecule is completely double-stranded.

After introduction, by transfection, into competent *E. coli* cells, DNA replication produces numerous copies of the recombinant DNA molecule. The semiconservative nature of DNA replication means that half the double-stranded molecules that are produced are unmutated in both strands, whereas half are mutated in both strands. Similarly, half the resulting phage progeny will carry copies of the unmutated molecule and half will carry the mutation. The phage produced by the transfected cells are plated on to solid agar so that plaques are produced. Half the plaques should contain the original recombinant molecule, and half the mutated version. Which are which is determined by plaque hybridization, using the oligonucleotide as the probe, and employing very strict conditions so that only the completely base-paired hybrid is stable.

(c) Studying the effect of the mutation Cells infected with M13 vectors do not lyse, but instead continue to divide (p. 20). A gene ligated into an M13 vector can therefore be expressed in the host cells resulting in production of recombinant protein. The protein coded by the cloned gene can be purified from the recombinant cells and its properties studied. The effect of a single base-pair mutation on the activity of the protein can therefore be assessed.

(d) The potential of oligonucleotide-directed mutagenesis

Oligonucleotide-directed mutagenesis and related techniques are proving to have remarkable potential, both for pure research and for applied biotechnology. For example, the biochemist can now ask very specific questions about the way that protein structure affects the action of an enzyme. In the past, it has been possible through biochemical analysis to gain some idea of the identity of the amino acids that provide the substrate binding and catalytic functions of an enzyme molecule. Mutagenesis techniques can provide a much more detailed picture by enabling the role of each individual amino acid to be assessed by replacing it with an alternative residue and determining the effect this has on the enzymatic activity. The ability to manipulate enzymes in this way has resulted in dramatic advances in our understanding of biological catalysis and led to the new field of **protein engineering**, in which

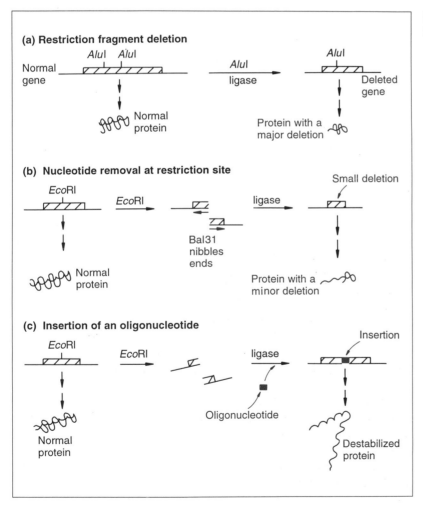

Figure 10.18 Various *in vitro* mutagenesis techniques.

Figure 10.19 One method for oligonucleotide-directed mutagenesis.

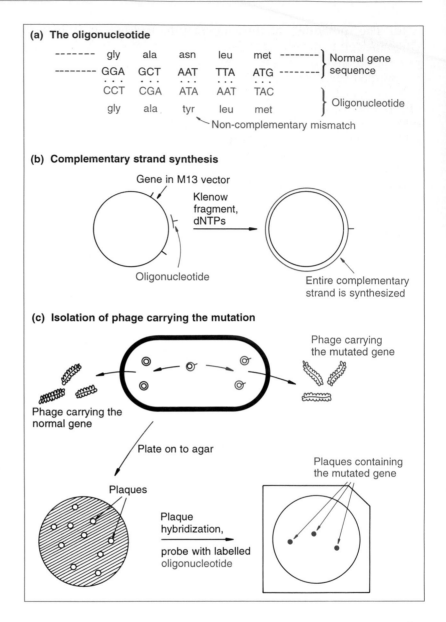

(a) The oligonucleotide

	gly	ala	asn	leu	met	
-------	gly	ala	asn	leu	met	-------- } Normal gene
--------	GGA	GCT	AAT	TTA	ATG	-------- } sequence
	CCT	CGA	ATA	AAT	TAC	
	gly	ala	tyr	leu	met	} Oligonucleotide

Non-complementary mismatch

(b) Complementary strand synthesis

Gene in M13 vector

Klenow fragment, dNTPs

Oligonucleotide

Entire complementary strand is synthesized

(c) Isolation of phage carrying the mutation

Phage carrying the mutated gene

Phage carrying the normal gene

Plate on to agar

Plaques

Plaque hybridization, probe with labelled oligonucleotide

Plaques containing the mutated gene

mutagenesis techniques are used to develop new enzymes for biotechnological purposes. For example, careful alterations to the amino acid sequence of subtilisin, an enzyme used in biological washing powders, have resulted in engineered versions with greater resistances to the thermal and bleaching (oxidative) stresses encountered in washing machines.

FURTHER READING

Favaloro, J. *et al.* (1980) Transcription maps of polyoma virus-specific RNA: analysis by two-dimensional nuclease S1 gel mapping. *Methods in Enzymology*, **65**, 718–49.

Galas, D. J. and Schmitz, A. (1978) DNase footprinting: a simple method for the detection of protein–DNA binding specificity. *Nucleic Acids Research*, **5**, 3157–70.

Fried, M. and Crothers, D. M. (1981) Equilibria and kinetics of *lac* repressor–operator interactions by polyacrylamide gel electrophoresis. *Nucleic Acids Research*, **9**, 6505–25 – gel retardation.

Garner, M. M. and Rezvin, A. (1981) A gel electrophoretic method for quantifying the binding of proteins to specific DNA regions: application to components of the *Escherichia coli* lactose operon regulatory system. *Nucleic Acids Research*, **9**, 3047–60 – gel retardation.

Paterson, B. M. *et al.* (1977) Structural gene identification and mapping by DNA.mRNA hybrid-arrested cell-free translation. *Proceedings of the National Academy of Sciences, USA*, **74**, 4370–4.

Smith, M. (1985) *In vitro* mutagenesis. *Annual Review of Genetics*, **19**, 423–62.

Kunkel, T. A. (1985) Rapid and efficient site-specific mutagenesis without phenotypic selection. *Proceedings of the National Academy of Sciences, USA*, **82**, 488–92 – oligonucleotide-directed mutagenesis.

Primrose, S. B. (1991) *Molecular Biotechnology*, 2nd edn, Blackwell Scientific Publications, Oxford – Chapter 4 gives a good overview of protein engineering.

11

The polymerase chain reaction

The development of gene cloning techniques in the 1970s provided a fresh impetus to research by enabling genes and gene activity to be studied in ways that previously had been impossible. Something very similar happened in the late 1980s when a second revolutionary procedure – the **polymerase chain reaction (PCR)** – was invented. PCR is a very uncomplicated technique: all that happens is that a short region of a DNA molecule, a single gene for instance, is copied many times by a DNA polymerase enzyme (see Figure 1.4). This may seem a rather trivial exercise but it has a multitude of applications in genetical research and in broader areas of biology.

We will begin this chapter with an outline of the polymerase chain reaction in order to understand exactly what it achieves. Then we will run through the relevant methodology, following the steps involved in PCR and the special methods that have been devised for studying the amplified DNA fragments that are obtained. Finally we will put PCR into context by surveying a few of the applications that have been found for this versatile technique.

11.1 THE POLYMERASE CHAIN REACTION IN OUTLINE

PCR results in the selective amplification of a chosen region of a DNA molecule. Any region of any DNA molecule can be chosen, so long as the sequences at the borders of the region are known. The border sequences must be known because in order to carry out a PCR, two short oligonucleotides must hybridize to the DNA molecule, one to each strand of the double helix (Figure 11.1). These oligonucleotides, which act as primers for the DNA synthesis reactions, delimit the region that will be amplified.

Amplification is usually carried out by the DNA polymerase I enzyme from *Thermus aquaticus*, the same bacterium that produces

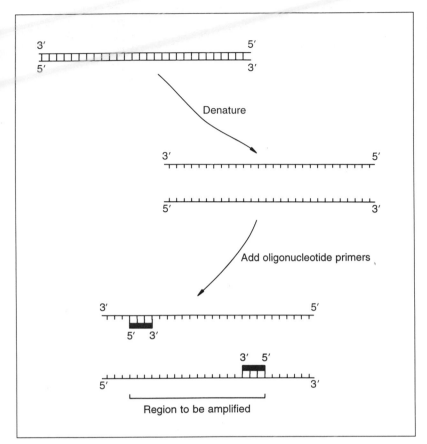

Figure 11.1 Hybridization of the oligonucleotide primers to the template DNA at the beginning of a PCR.

the restriction enzyme (*Taq*I (p. 68). This organism lives in hot springs, and many of its enzymes, including the *Taq* polymerase, are thermostable, meaning that they are resistant to denaturation by heat treatment. As will be apparent in a moment, the thermostability of the *Taq* polymerase is an essential requirement in PCR methodology.

To begin a PCR amplification, the enzyme is added to the primed template DNA and incubated so that it synthesizes new complementary strands (Figure 11.2(a)). The mixture is then heated to 94°C so that the newly synthesized strands detach from the template (Figure 11.2(b)), and cooled, enabling more primers to hybridize at their respective positions, including positions on the newly synthesized strands. The *Taq* polymerase, which unlike most types of DNA polymerase is not inactivated by the heat treatment, now carries out a second round of DNA synthesis (Figure 11.2(c)). The cycle of denaturation–hybridization–synthesis is repeated, usually 25–30 times, resulting in the eventual synthesis of several hundred million copies of the amplified DNA fragment (Figure 11.2(d)).

Figure 11.2 The polymerase chain reaction.

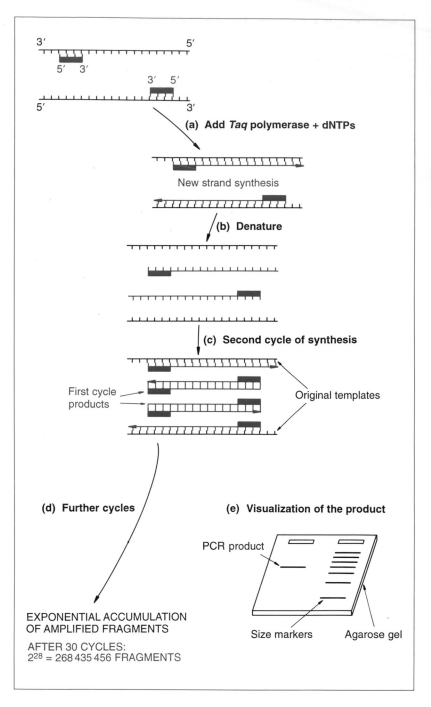

(a) **Add *Taq* polymerase + dNTPs**

New strand synthesis

(b) **Denature**

(c) **Second cycle of synthesis**

First cycle products

Original templates

(d) **Further cycles**

(e) **Visualization of the product**

PCR product

Size markers   Agarose gel

EXPONENTIAL ACCUMULATION
OF AMPLIFIED FRAGMENTS

AFTER 30 CYCLES:
$2^{28} = 268\,435\,456$ FRAGMENTS

At the end of a PCR experiment a sample of the reaction mixture is usually analysed by agarose gel electrophoresis, sufficient DNA having been produced for the amplified fragment to be visible as a discrete band after staining with ethidium bromide (Figure

11.2(e)). As will be described later in this chapter, this may by itself provide useful information about the DNA region that has been amplified. Alternatively the PCR product can be ligated into a plasmid or bacteriophage vector, cloned in the normal way, and examined by standard techniques such as DNA sequencing.

11.2 PCR IN MORE DETAIL

Although PCR experiments are very easy to set up, they must be planned carefully if the results are to be of any value. The sequences of the primers are critical to the success of the experiment, as are the precise temperatures used in the heating and cooling stages of the reaction cycle. Also there is the important question of what can be done with the amplified DNA molecules once they have been obtained.

11.2.1 Designing the oligonucleotide primers for a PCR

The primers are the key to the success or failure of a PCR experiment. If the primers are designed correctly then the experiment will result in amplification of a single DNA fragment, corresponding to the target region of the template molecule. If the primers are incorrectly designed then the experiment will fail, possibly because no amplification occurs, or possibly because the wrong fragment, or more than one fragment, is amplified (Figure 11.3). Clearly a great deal of thought must be put into the design of the primers.

Working out appropriate sequences for the primers is not a problem: they must correspond with the sequences flanking the target region on the template molecule. Each primer must of course be complementary (not identical) to its template strand in order for

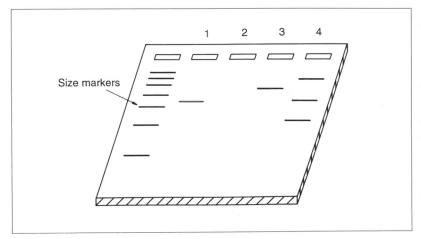

Figure 11.3 The results of PCRs with well-designed and poorly-designed primers. Lane 1 shows a single amplified fragment of the expected size, the result of a well-designed experiment. In Lane 2 there is no amplification product, suggesting that one or both of the primers were unable to hybridize to the template DNA. Lanes 3 and 4 show, respectively, an amplification product of the wrong size, and a mixture of products (the correct product plus two wrong ones); both results are due to hybridization of one or both of the primers to non-target sites on the template DNA molecule.

hybridization to occur, and the 3′ ends of the hybridized primers should point towards one another (Figure 11.4). The DNA fragment to be amplified should not be greater than about 3 kb in length and ideally will be less than 1 kb. Fragments up to 10 kb can be amplified by standard PCR techniques, but the longer the fragment the less efficient the amplification and the more difficult it is to obtain consistent results. Amplification of very long fragments – up to 40 kb – is possible but requires special methods.

Figure 11.4 A pair of primers designed to amplify the human α1-globin gene. The exons of the gene are shown as closed boxes, the introns as open boxes.

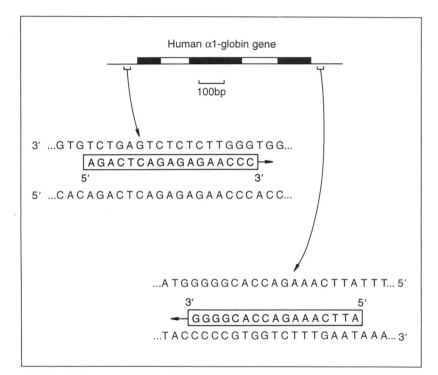

Now the important issue arises: how long should the primers be? If the primers are too short then they might hybridize to non-target sites and give undesired amplification products. To illustrate this point, imagine that total human DNA is used in a PCR experiment with a pair of primers 8 nucleotides in length (in PCR jargon these are called '8-mers'). The likely result (Figure 11.5(a)) is that a number of different fragments will be amplified. This is because attachment sites for these primers are expected to occur, on average, once every $4^8 = 65\,536$ bp, giving approximately 46 000 possible sites in the 3 000 000 kb of nucleotide sequence that makes up the human genome. This means that it would be very unlikely that a pair of 8-mer primers would give a single, specific amplification product with human DNA.

What if the 17-mer primers shown in Figure 11.4 are used? The

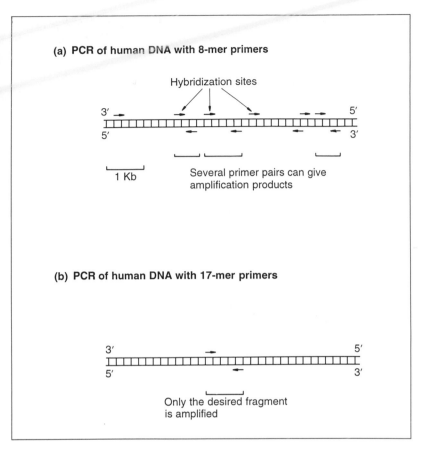

Figure content labels:

(a) **PCR of human DNA with 8-mer primers**

Hybridization sites

3′ 5′

5′ 3′

1 Kb

Several primer pairs can give amplification products

(b) **PCR of human DNA with 17-mer primers**

3′ 5′

5′ 3′

Only the desired fragment is amplified

Figure 11.5 The lengths of the primers are critical for the specificity of the PCR.

expected frequency of a 17-mer sequence is once every 4^{17} = 17 179 869 184 bp. This figure is over five times greater than the length of the human genome, so a 17-mer primer would be expected to have just one hybridization site in total human DNA. A pair of 17-mer primers should therefore give a single, specific amplification product (Figure 11.5(b)).

Why not simply make the primers as long as possible? Because the length of the primer influences the rate at which it hybridizes to the template DNA, longer primers hybridizing at a slower rate. The efficiency of the PCR, measured by the number of amplified molecules produced during the experiment, is therefore reduced if the primers are too long, as complete hybridization to the template molecules can not occur in the time allowed during the reaction cycle. In practice, primers longer than 30-mer are rarely used.

11.2.2 Working out the correct temperatures to use

During each cycle of a PCR the reaction mixture is transferred between three temperatures (Figure 11.6):

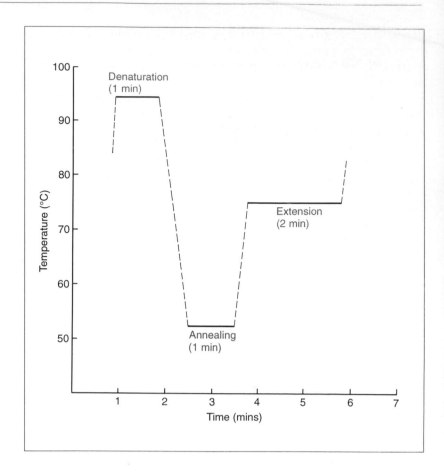

Figure 11.6 A typical temperature profile for a PCR.

1. The denaturation temperature, usually 94°C, which breaks the base pairs and releases single-stranded DNA to act as templates in the next round of DNA synthesis.
2. The hybridization or annealing temperature, at which the primers attach to the templates.
3. The extension temperature, at which DNA synthesis occurs. This is usually set at 74°C, just below the optimum for *Taq* polymerase.

The annealing temperature is the important one as again this can affect the specificity of the reaction. DNA–DNA hybridization is a temperature-dependent phenomenon. If the temperature is too high then no hybridization takes place, instead the primers and templates remain dissociated (Figure 11.7(a)). However, if the temperature is too low then mismatched hybrids – ones in which not all the correct base pairs have formed – are stable (Figure 11.7(b)). If this occurs then the earlier calculations regarding the appropriate lengths for the primers become irrelevant, as the calculations assumed that only perfect primer–template hybrids are able to form. If mismatches are tolerated then the number of potential hybridization sites for each primer is greatly increased, and amplification is more likely to occur at non-target sites on the template molecule.

The ideal annealing temperature must be low enough to enable hybridization between primer and template, but high enough to prevent mismatched hybrids from forming (Figure 11.7(c)). This temperature can be estimated by determining the **melting temperature** or T_m of the primer–template hybrid. The T_m is the temperature at which the correctly base-paired hybrid dissociates ('melts'): a temperature 1–2°C below this should be low enough to allow the correct primer–template hybrid to form, but too high for a hybrid

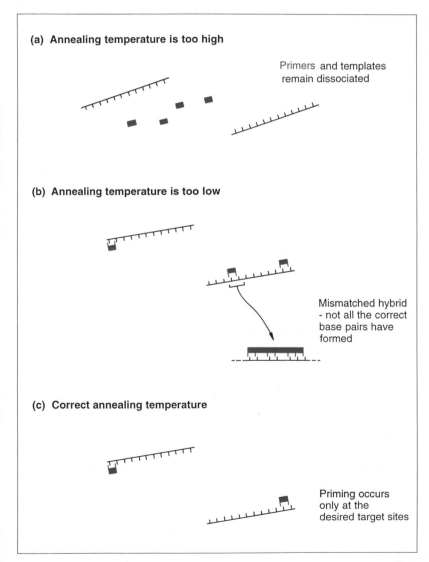

(a) Annealing temperature is too high

Primers and templates remain dissociated

(b) Annealing temperature is too low

Mismatched hybrid - not all the correct base pairs have formed

(c) Correct annealing temperature

Priming occurs only at the desired target sites

Figure 11.7 Temperature has an important effect on the hybridization of the primers to the template DNA.

Figure 11.8 Calculating the T_m of a primer.

Primer sequence: 5′ AGACTCAGAGAGAACCC 3′

4Gs   5Cs   7As   1T

T_m = (4 x 9) + (2 x 8)

= 36 + 16

= 52°C

with a single mismatch to be stable. The T_m can be determined experimentally but is more usually calculated from the simple formula (Figure 11.8):

$$T_m = (4 \times [G+C]) + (2 \times [A+T]) \,°C$$

in which $[G+C]$ is the number of G and C nucleotides in the primer sequence, and $[A+T]$ is the number of A and T nucleotides.

The annealing temperature for a PCR experiment is therefore determined by calculating the T_m for each primer and using a temperature 1–2°C below this figure. Note that this means that the two primers should be designed so that they have identical T_ms. If this is not the case then the appropriate annealing temperature for one primer may be too high or too low for the other member of the pair.

11.2.3 After the PCR: studying PCR products

PCR is often the starting point for a longer series of experiments in which the amplification product is studied in various ways in order to gain information about the DNA molecule that acted as the original template. Although a wide range of techniques have been devised for studying PCR products the three most important types of analysis are as follows:

1. Agarose gel electrophoresis.
2. Direct sequence analysis of the PCR product.
3. Analyses that involve cloning the PCR product.

(a) Gel electrophoresis of PCR products With most PCR experiments, the results are checked by running a portion of the amplified reaction mixture in an agarose gel. A band representing the amplified DNA may be visible after ethidium bromide staining, or if the DNA yield is low the product can be detected by Southern hybridization (p. 184). If the expected band is absent, or if additional bands are present, then something has gone wrong and the experiment must be repeated.

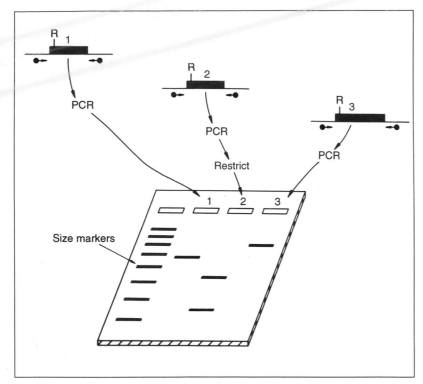

Figure 11.9 Gel electrophoresis of the PCR product can provide information on the template DNA molecule. Lanes 1 and 2 show, respectively, an unrestricted PCR product and a product restricted with the enzyme that cuts at site 'R'. Lane 3 shows the result obtained when the template DNA contains an insertion in the amplified region.

In some cases, agarose gel electrophoresis is used not only to determine if the PCR experiment has worked, but also to obtain additional information. For example, the presence of restriction sites in the amplified region of the template DNA can be determined by treating the PCR product with a restriction endonuclease before running the sample in the agarose gel. Alternatively, the exact size of the PCR product can be used to establish if the template DNA contains an insertion or deletion mutation in the amplified region (Figure 11.9). In both cases the PCR experiment is used as a modified form of RFLP analysis (p. 200), with the advantage over the traditional system that very little starting material is needed, making the PCR-based approach more suitable for applications such as prenatal screening.

(b) Direct sequence analysis of PCR products   If gel electrophoresis on its own is not sufficient to provide useful information then often the next step is to determine the sequence of the amplified DNA fragment. This can be achieved by cloning the PCR product (see below) and using a standard sequencing method (p. 192), or alternatively by sequencing the PCR product directly.

The direct sequencing method is based on the Sanger–Coulson technique (p. 193) and therefore requires single-stranded DNA as the starting material. The PCR product is of course double-stranded, so some means of purifying single strands is needed. There are several possibilities, the best of these being to carry out the initial PCR with one normal and one modified primer, the modified primer altered in such a way that the DNA strands synthesized from it are easily purified. A clever way of doing this is by attaching small magnetic beads to one of the primers. After the PCR, single-stranded DNA is obtained by separating the 'magnetic' strand from the ordinary strand (Figure 11.10). A similar method makes use of a biotin-labelled primer, with the single strands separated by binding to avidin, a protein that has a high affinity for biotin (p. 172).

Once single-stranded DNA has been purified, the remainder of the procedure is similar to the standard Sanger–Coulson method, in which families of chain-terminated molecules are synthesized by a DNA polymerase such as Sequenase. The only complication concerns the sequencing primer used as the starting point for the strand synthesis reactions. In the standard sequencing procedure this primer anneals to a site within the M13 vector, adjacent to the polylinker into which the DNA to be sequenced is cloned (see Figure 9.7). This 'universal' primer cannot be used with PCR products as these do not contain the appropriate M13 sequences. Instead one of the primers used for the initial PCR is also used to prime the sequencing reactions. This primer has to be complementary to the purified single strands, and therefore is the primer that was not labelled with magnetic beads or with biotin.

(c) Cloning PCR products If even more information is required about the PCR product then the fragments can be ligated into a vector and examined by any of the standard methods for studying cloned DNA (see Chapters 9 and 10). This may sound easy but there are complications.

The first problem concerns the ends of the PCR products. From an examination of Figure 11.2 it might be imagined that fragments amplified by PCR are blunt-ended. If this was the case then they could be inserted into a cloning vector by blunt-end ligation, or alternatively the PCR products could be provided with sticky ends by the attachment of linkers or adaptors (p. 79). Unfortunately, the situation is not so straightforward. *Taq* polymerase tends to add an additional nucleotide, usually an adenosine, to the end of each strand that it synthesizes. This means that a double-stranded PCR product is not blunt-ended, instead each 3' terminus has a single nucleotide overhang (Figure 11.11). The overhangs could be removed by treatment with an exonuclease enzyme, resulting in PCR products with true blunt ends, but this is not a popular approach as it is difficult to prevent the exonuclease from becoming overactive and causing further damage to the ends of the molecules.

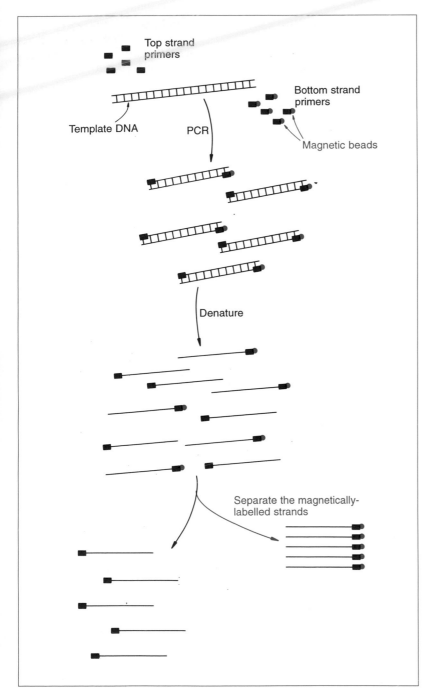

Top strand
primers

Bottom strand
primers

Template DNA

PCR

Magnetic beads

Denature

Separate the magnetically-
labelled strands

Figure 11.10 One way of purifying single-stranded DNA from a double-stranded PCR product. One of the primers is labelled with a magnetic bead. After PCR, the double-stranded products are denatured and the magnetic strands separated from the unlabelled strands.

One solution is to use a special cloning vector which carries T overhangs and which can therefore be ligated to an amplified DNA fragment (Figure 11.12). These vectors are usually prepared by restricting a standard vector at a blunt-end site, and then treating

Figure 11.11 Polynucleotides synthesized by *Taq* polymerase usually have an extra adenosine at their 3′ ends.

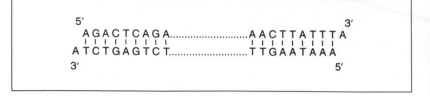

with *Taq* polymerase in the presence of just dTTP. No primer is present so all the polymerase can do is add a T nucleotide to the 3′ ends of the blunt-ended vector molecule, resulting in the T-tailed vector into which the PCR products can be inserted.

Figure 11.12 Using a special T-tailed vector to clone a PCR product.

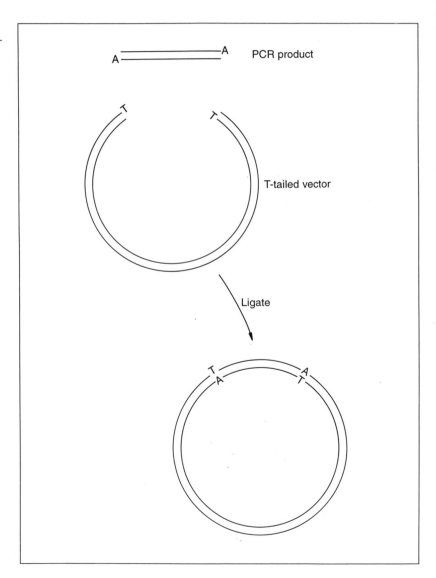

A second, more popular solution is to design primers that contain restriction sites. After PCR the amplified fragments are treated with the restriction endonuclease, which cuts each molecule within the primer sequence, leaving sticky-ended fragments that can be ligated efficiently into a standard cloning vector (Figure 11.13). The approach is not limited to those instances when the primers span restriction sites that are present in the template DNA. Instead, restriction sites can be added to the primer sequences as short segments at each 5′ end (Figure 11.14). These segments can not hybridize to the template molecule, but they are copied during the PCR reaction, resulting in PCR products that carry terminal restriction sites.

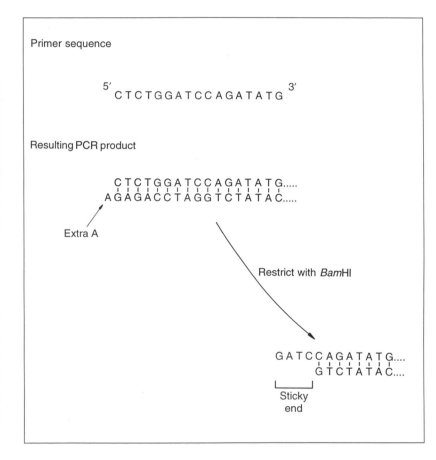

Figure 11.13 Obtaining a PCR product with a sticky end through use of a primer whose sequence includes a restriction site.

(d) Problems with the error rate of *Taq* polymerase   All DNA polymerases make mistakes during DNA synthesis, occasionally adding an incorrect nucleotide to the growing DNA strand. Most polymerases, however, are able to rectify these errors by reversing over the mistake and resynthesizing the correct sequence. This

Figure 11.14 A PCR primer with a restriction site present within an extension at the 5′ end.

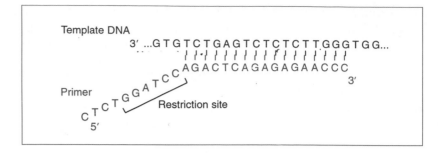

property is referred to as the 'proofreading' function and depends on the polymerase possessing a 3′ to 5′ exonuclease activity (p. 53).

Taq polymerase appears to lack a proofreading activity and as a result is unable to correct its errors. This means that the DNA synthesized by *Taq* polymerase is not always an accurate copy of the template molecule. The error rate has been estimated at one mistake for every 9000 nucleotides of DNA that is synthesized, which might appear to be almost insignificant but which translates to one error per 300 bp for the amplification products obtained after 30 cycles. This is because PCR involves copies being made of copies of copies, so the polymerase-induced errors gradually accumulate, the fragments produced at the end of a PCR containing copies of earlier errors along with any new errors introduced during the final round of synthesis.

For many applications this high error rate does not present a problem. In particular, direct sequencing of a PCR product (p. 237) provides the correct sequence of the template, even though the PCR products contain the errors introduced by the *Taq* polymerase. This is because the errors are distributed randomly, so for every amplified fragment that has an error at a particular nucleotide position there will be many molecules with the correct sequence. In this context the error rate is indeed insignificant.

This is not the case if the PCR products are cloned. Each resulting clone contains multiple copies of a single amplified fragment, so the cloned DNA does not necessarily have the same sequence as the original template molecule used in the PCR (Figure 11.15). This possibility lends an uncertainty to all experiments carried out with cloned PCR products and dictates that, whenever possible, the amplified DNA should be studied directly rather than being cloned.

11.3 APPLICATIONS OF PCR

At first it might be difficult to understand why PCR is such a useful technique. All that it achieves is amplification of a predeter-

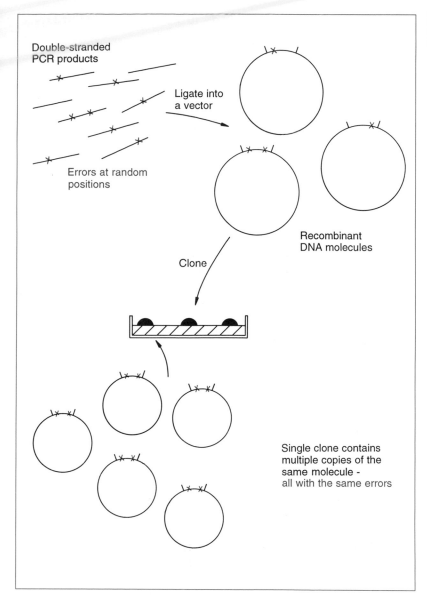

Double-stranded
PCR products

Errors at random
positions

Ligate into
a vector

Recombinant
DNA molecules

Clone

Single clone contains
multiple copies of the
same molecule -
all with the same errors

Figure 11.15 The high error rate of the *Taq* polymerase becomes a factor when PCR products are cloned.

mined fragment of DNA, with the apparent disadvantage that the border sequences must be known, the latter precluding PCR from analysis of DNA regions that have not previously been studied by standard methods. If PCR is unable to provide information on novel regions of genomes then why has it become so important in molecular biology?

There are many reasons, a few of which are described below.

11.3.1 PCR can be used to study minute quantities of DNA

PCR has a proven ability to use a single DNA molecule as the template for an amplification reaction. This has been demonstrated by the successful amplification of DNA from an isolated sperm cell containing just one, haploid copy of the human genome. This amazing sensitivity means that molecular biological analysis can now be applied to specimens that do not contain enough DNA for standard cloning procedures. This aspect of PCR is proving important in forensic analysis, enabling genetic fingerprinting techniques (p. 202) to be used with single hairs and even bloodstains, making it even more difficult for criminals to escape the long arm of the law. PCR has also been used to amplify DNA from the bones of murder victims, in some cases allowing identifications to be made with remains that are too badly decayed for conventional analysis. A notable example of this has been in establishing that the bones recovered from Ekaterinburg do indeed belong to Tsar Nicholas II and his family.

The extreme sensitivity of PCR has also opened up new possibilities in archaeology and palaeontology, by enabling nucleotide sequences to be obtained from traces of DNA present in preserved or fossilized material. PCR analysis has been used to study the genetic affinities of ancient peoples through amplification and sequence analysis of traces of DNA retained in their bones or preserved in specimens such as mummies and bog bodies. The origins of the first Americans and the migration routes that led to colonization of the Pacific are already being unravelled, and DNA analysis taken back into geological time with amplification of sequences from fossilized plant remains and insects embedded in amber.

11.3.2 PCR can be used in clinical diagnosis

Standard RFLP analysis is extensively used to screen for human gene mutations that might lead to genetic disease, but is applicable only in those cases where the mutation results in a detectable change in the length of a restriction fragment (p. 200). Mutations for many genetic diseases do not result in an RFLP and can only be detected if the relevant region of the genome is sequenced. With conventional techniques this means preparing a gene library for each individual, and then isolating and sequencing the clone containing the mutated gene (Figure 11.16(a)). Although this approach is by no means impossible it is too time-consuming to be used within a routine screening programme. PCR provides a feasible alternative by enabling sequence information to be obtained very quickly by amplification of the relevant region of the genome followed by direct analysis of the PCR products (Figure 11.16(b)). This ability to type mutations rapidly is not only important in clinical diagnosis, it has also accelerated research into genetic disease by allowing many different at-risk groups to be examined so that novel mutations can be identified.

The sensitivity of PCR has also led to applications in the diagnosis of pathogenic disease. Amplification of, for example, viral DNA in human samples often enables diagnosis to be made days, weeks or even months before the onset of symptoms. The relevance is of course that the treatment of many diseases, especially the virally-induced cancers (e.g. cervical cancer, caused by human

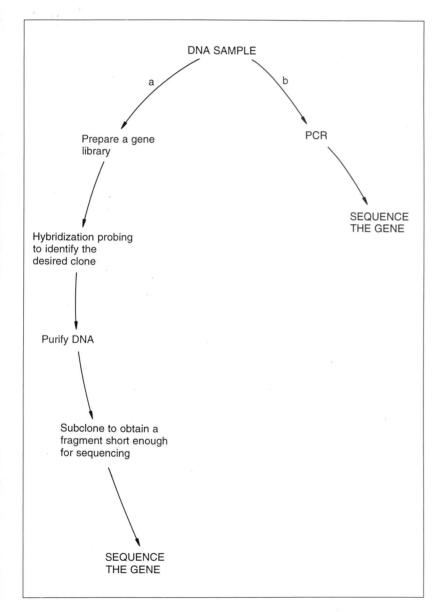

Figure 11.16 A comparison between the (a) traditional and (b) PCR-based methods for obtaining the sequence of a gene from human DNA.

papillomaviruses), is generally more successful if begun at the earliest possible stage.

11.3.3 PCR can be used to amplify RNA

PCR is not limited to the amplification of DNA templates. RNA molecules can also be amplified if they are first converted into single-stranded cDNA with the enzyme reverse transcriptase (p. 58). Once this preliminary step has been carried out the PCR primers and *Taq* polymerase are added and the experiment proceeds exactly as in the standard technique (Figure 11.17).

Figure 11.17 Reverse transcription–PCR (RT–PCR).

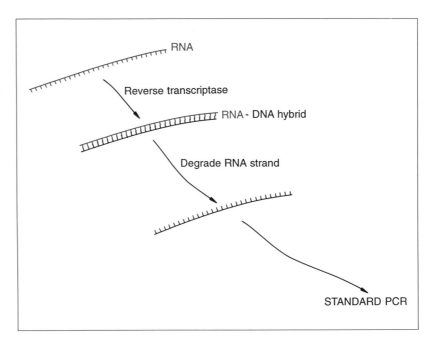

A useful application of **RT–PCR** is in measuring the relative amounts of an mRNA in different tissues or in the same tissue at different times. The amount of an mRNA in a cell is generally taken to be a reflection of the activity of the parent gene, so quantification of the mRNA enables changes in gene expression to be monitored. In the past, quantification has been carried out by northern hybridization of RNA extracts (p. 184), but this is possible only for those mRNAs that are relatively abundant, rare RNAs being undetectable by hybridization techniques. PCR, being much more sensitive, allows expression studies to be extended to the less active genes, which are often the most interesting ones.

PCR methods for quantification are based on the assumption that the amount of product that is synthesized is proportional to the amount of template RNA (or DNA) present at the start of the reaction. The relative amounts of product from two PCRs can be estimated by comparing the intensities of the bands seen when samples are electrophoresed in an agarose gel (Figure 11.18(a)). An estimate of the actual amount of mRNA present at the start of the PCR can be made by comparing the band intensity with a series of controls produced by amplifying known amounts of DNA (Figure 11.18(b)). This is most accurate if the DNA amplifications are carried out in the same tubes as the RT–PCRs, using the same primers. For this to be possible the mRNA must be derived from a gene

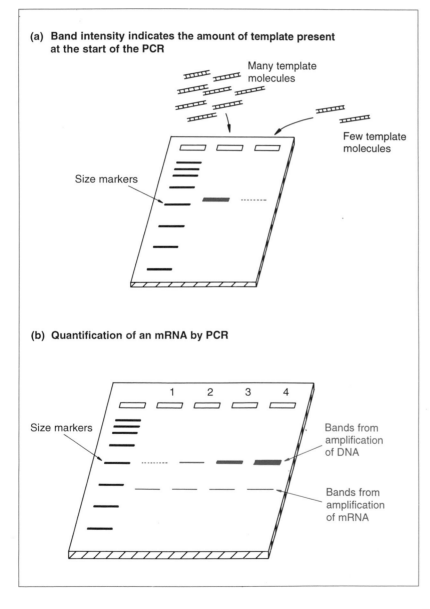

(a) Band intensity indicates the amount of template present at the start of the PCR

Many template molecules

Few template molecules

Size markers

(b) Quantification of an mRNA by PCR

1 2 3 4

Size markers

Bands from amplification of DNA

Bands from amplification of mRNA

Figure 11.18 Quantification of the products of RT–PCR. In (b), PCRs have been carried out with samples containing equal amounts of RNA but increasing amounts of DNA. The DNA contains the gene from which the mRNA is transcribed, and the amplified fragment spans an intron within the gene. The PCR product derived from the DNA is therefore longer than that from the mRNA. The RNA- and DNA-derived bands have similar intensities in Lane 2, indicating that this PCR contained an approximately equal number of copies of the DNA and mRNA versions of the target sequence.

containing one or more introns, so that the amplified fragments from the DNA template are longer than the RNA-derived products and hence migrate to a different position in the agarose gel.

11.3.4 PCR can be used to compare different genomes

As illustrated in Figure 11.5(a), if the primers used in a PCR are too short then a mixture of amplified fragments will be obtained. From the discussion on p. 232 it might be imagined that this situation should be avoided at all costs. Random amplification with short primers is, however, a useful technique in phylogenetics, the area of research concerned with the evolutionary history and lines of descent of species and other groups of organisms. The important point is that the banding pattern seen when the products of PCR with random primers are electrophoresed is a reflection of the overall structure of the DNA molecule used as the template. If the starting material is total cell DNA then the banding pattern represents the organization of the cell's genome. Differences between the genomes of two organisms, whether members of the same or of different species, can therefore be measured by PCR with random primers. Two closely related organisms would be expected to yield more similar banding patterns than two organisms that are more distant in evolutionary terms (Figure 11.19). The technique is referred to as **random amplified polymorphic DNA (RAPD) analysis**.

As with many phylogenetic techniques, the interpretation of RAPD analysis is highly complex and as yet there is no agreement regarding the way in which the data should be handled. Despite

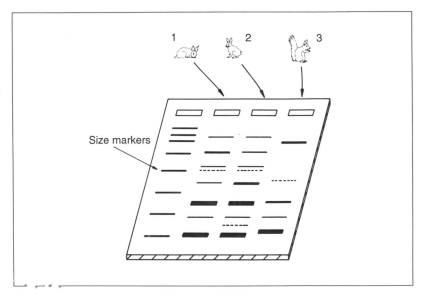

Figure 11.19 RAPD analysis. A research group decides to test the hypothesis that a squirrel is a rabbit with an unusual tail and a predisposition for nuts. PCRs with random primers are carried out with DNA from two rabbits (Lanes 1 and 2) and a squirrel (Lane 3). The two rabbits give very similar banding patterns but the squirrel gives a different pattern. This shows that a squirrel is different from a rabbit; the research group has to form a new hypothesis.

this problem a number of interesting experiments have already been carried out. In one project it was shown that fruiting bodies of the honey mushroom (*Armillaria bulbosa*) collected from a large site in northern Michigan all had identical genome structures, not even displaying the small differences expected for individuals within a single species. The interpretation is that the site contains just a single, giant clone of the fungus, covering 37 acres and probably one of the oldest organisms on the planet.

FURTHER READING

Saiki, R. K., Gelfand, D. H., Stoffel, S., Scharf, S. J., Higuchi, R., Horn, G. T., Mullis, K. B. and Erlich, H. A. (1988) Primer-directed enzymatic amplification of DNA with a thermostable DNA polymerase. *Science*, **239**, 487–91 – the first description of PCR with *Taq* polymerase.

Rychlik, W., Spencer, W. J. and Rhoads, R. E. (1990) Optimization of the annealing temperature for DNA amplification *in vitro*. *Nucleic Acids Research*, **18**, 6409–12.

Marchuk, D., Drumm, M., Saulino, A. and Collins, F. S. (1991) Construction of T-vectors, a rapid and general system for direct cloning of unmodified PCR products. *Nucleic Acids Research*, **19**, 1154.

Gill, P., Ivanov, P. L., Kimpton, C., Piercy, R., Benson, N., Tully, G., Evett, I., Hagelberg, E. and Sullivan, K. (1994) Identification of the remains of the Romanov family by DNA analysis. *Nature Genetics*, **6**, 130–5.

Brown, T. A. and Brown, K. A. (1994) Ancient DNA: using molecular biology to study the past. *BioEssays*, **16**, 719–26.

Foley, K. P., Leonard, M. W. and Engel, J. D. (1993) Quantitation of RNA using the polymerase chain reaction. *Trends in Genetics*, **9**, 380–5.

Smith, M. L., Bruhn, J. N. and Anderson, J. B. (1992) The fungus *Armillaria bulbosa* is among the largest and oldest living organisms. *Nature*, **356**, 428–31 – an example of the use of RAPD analysis.

Part Three
Gene Cloning in Research and Biotechnology

Production of protein from cloned genes 12

Now that we have covered the basic techniques involved in cloning a gene and studying its structure and expression, we can move on to consider how recombinant DNA technology is being applied in biological research. This will lead us into one of the major growth industries of the late twentieth century – biotechnology.

Biotechnology is not a new subject, although it has received far more attention during the last few years than it ever has in the past. Biotechnology can be defined as the use of living organisms in industrial or industrial-type processes. According to archaeologists, the British biotechnology industry dates back 4000 years, to the late Neolithic period, when fermentation processes that make use of living yeast cells to produce ale and mead were first introduced into this country. Certainly brewing was well established by the time of the Roman invasion.

During the twentieth century, biotechnology has expanded with the development of a variety of industrial uses for microorganisms. The discovery by Alexander Fleming in 1929 that the mould *Penicillium* synthesizes a potent antibacterial agent led to the use of fungi and bacteria in the large-scale production of antibiotics. At first the microorganisms were grown in large culture vessels from which the antibiotic was purified after the cells had been removed (Figure 12.1(a)), but more recently this **batch culture** method has been largely supplanted by **continuous culture** techniques, making use of a **fermenter**, from which samples of medium can be continuously drawn off, providing a non-stop supply of the product (Figure 12.1(b)). This type of process is not limited to antibiotic production and has also been used to obtain large amounts of other compounds produced by microorganisms (Table 12.1).

One of the reasons why biotechnology has received so much attention during the last decade is because of gene cloning. Although many useful products can be obtained from microbial culture, the list in the past has been limited to those compounds

Figure 12.1 Two different systems for the growth of microorganisms. (a) Batch culture. (b) Continuous culture.

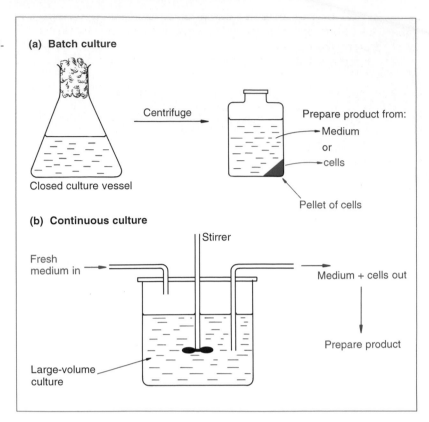

naturally synthesized by microorganisms. Many important pharmaceuticals that are produced not by microbes but by higher organisms could not be obtained in this way. This has been changed by the application of gene cloning to biotechnology. The ability to clone genes means that a gene for an important animal or plant protein can now be taken from its normal host, inserted into a cloning vector, and introduced into a bacterium (Figure 1.5). If the manipulations are performed correctly then the gene will be expressed and the protein synthesized by the bacterial cell. It may then be possible to obtain large amounts of the protein.

Of course, in practice obtaining **recombinant protein** is not as easy as it sounds. Special types of cloning vector are needed, and satisfactory yields of recombinant protein are often difficult to obtain. In this chapter we will look at cloning vectors for recombinant protein synthesis and examine some of the problems associated with their use.

Table 12.1   Some of the compounds produced by industrial-scale culture of microorganisms

Compound	Microorganism
Antibiotics	
Penicillins	*Penicillium* spp.
Cephalosporins	*Cephalosporium* spp.
Gramicidins, polymixins	*Bacillus* spp.
Chloramphenicol, streptomycin	*Streptomyces* spp.
Enzymes	
Invertase	*Saccharomyces cerevisiae*
Proteases, amylases	*Bacillus* spp., *Aspergillus* spp.
Others	
Alcohol	*S. cerevisiae, S. carlsbergensis*
Glycerol	*S. cerevisiae*
Vinegar	*S. cerevisiae*, acetic acid bacteria
Dextran	*Leuconostoc* spp.
Butyric acid	Butyric acid bacteria
Acetone, butanol	*Clostridium* spp.
Citric acid	*Aspergillus niger*

12.1 SPECIAL VECTORS FOR EXPRESSION OF FOREIGN GENES IN *E. COLI*

If a foreign (i.e. non-bacterial) gene is simply ligated into a standard vector and cloned in *E. coli* then it is very unlikely that a significant amount of recombinant protein will be synthesized. This is because expression is dependent on the gene being surrounded by a collection of signals that can be recognized by the bacterium. These signals, which are short sequences of nucleotides, advertise the presence of the gene and provide instructions for the transcriptional and translational apparatus of the cell. The three most important signals for *E. coli* genes are as follows (Figure 12.2):

1. The **promoter**, which marks the point at which transcription of the gene should start. In *E. coli* the promoter is recognized by the sigma subunit of the transcribing enzyme RNA polymerase.
2. The **terminator**, which marks the point at the end of the gene where transcription should stop. A terminator is usually a nucleotide sequence that can base-pair with itself to form a **stem-loop** structure.
3. The **ribosome binding site**, a short nucleotide sequence recognized by the ribosome as the point at which it should attach to

Figure 12.2 The three most important signals for gene expression in
E. coli.

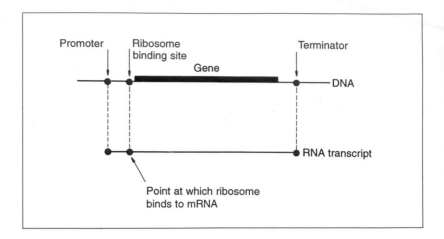

the mRNA molecule. The initiation codon of the gene is always
a few nucleotides downstream of this site.

The genes of higher organisms are also surrounded by expression
signals, but their nucleotide sequences are not the same as the
E. coli versions. This is illustrated by comparing the promoters of
E. coli and human genes (Figure 12.3). There are similarities, but it
is unlikely that an *E. coli* RNA polymerase would be able to attach
to a human promoter. A foreign gene is inactive in *E. coli* quite simply because the bacterium does not recognize its expression signals.

A solution to this problem would be to insert the foreign gene
into the vector in such a way that it is placed under control of a set
of *E. coli* expression signals. If this can be achieved then the gene
should be transcribed and translated (Figure 12.4). Cloning vehicles which provide these signals, and which can therefore be used
in the production of recombinant protein, are called **expression
vectors**.

Figure 12.3 Typical promoter
sequences for *E. coli* and animal
genes.

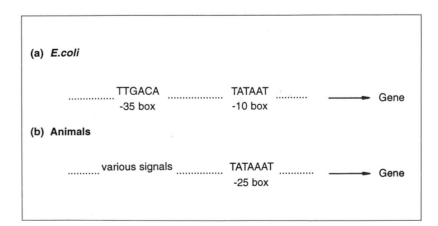

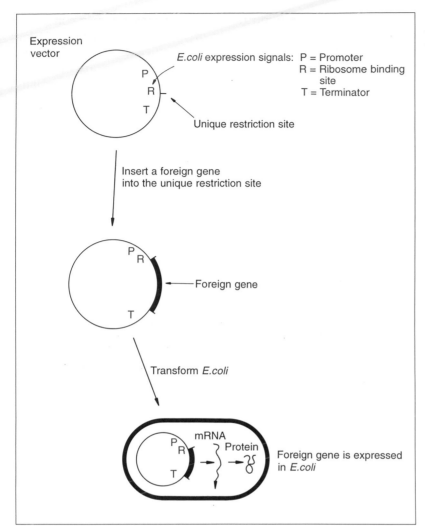

Figure 12.4 The use of an expression vector to achieve expression of a foreign gene in *E. coli*.

12.1.1 The promoter is the critical component of an expression vector

The promoter is the most important component of an expression vector. This is because the promoter controls in the very first stage of gene expression (attachment of an RNA polymerase enzyme to the DNA) and determines the rate at which mRNA is synthesized. The amount of recombinant protein that is obtained therefore depends to a great extent on the nature of the promoter carried by the expression vector.

(a) The promoter must be chosen with care   The two sequences shown in Figure 12.3(a) are **consensus sequences**, averages of all the *E. coli* promoter sequences that are known. Although most *E. coli* promoters do not differ much from these consensus sequences (e.g. TTTACA instead of TTGACA), a small variation may in fact have a major effect on the efficiency with which the promoter can direct transcription. **Strong promoters** are those that can sustain a high rate of transcription; strong promoters usually control genes whose translation products are required in large amounts by the cell (Figure 12.5(a)). In contrast, **weak promoters**, which are relatively inefficient, direct transcription of genes whose products are needed in only small amounts (Figure 12.5(b)). Clearly an expression vector should carry a strong promoter, so that the cloned gene is transcribed at the highest possible rate.

Figure 12.5 Strong and weak promoters.

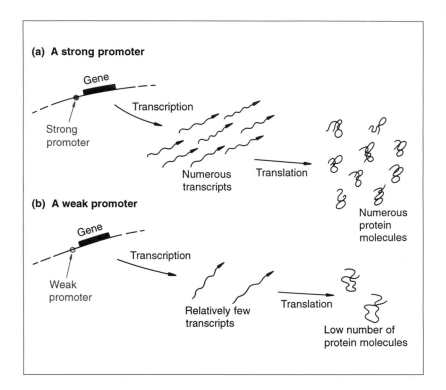

(a) A strong promoter

Gene

Strong promoter

Transcription

Numerous transcripts

Translation

Numerous protein molecules

(b) A weak promoter

Gene

Weak promoter

Transcription

Relatively few transcripts

Translation

Low number of protein molecules

A second factor to be considered when constructing an expression vector is whether it will be possible to regulate the promoter in any way. Two major types of gene regulation are recognized in *E. coli* – **induction** and **repression**. An inducible gene is one whose transcription is switched on by addition of a chemical to the growth medium; often this chemical is one of the substrates for the enzyme coded by the inducible gene (Figure 12.6(a)). In contrast, a repressible gene is switched off by addition of the regulatory chemical (Figure 12.6(b)).

Gene regulation is a complex process that only indirectly involves the promoter itself. However, many of the sequences

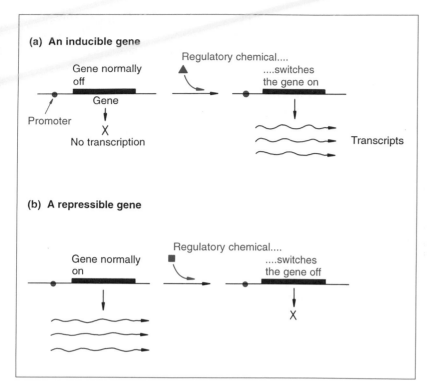

Figure 12.6 Examples of the two major types of gene regulation that occur in bacteria. (a) An inducible gene. (b) A repressible gene.

important in induction and repression lie in the region surrounding the promoter and are therefore also present in an expression vector. It may therefore be possible to extend the regulation to the expression vector, so that the chemical that induces or represses the gene normally controlled by the promoter is also able to regulate expression of the cloned gene.

This can be a distinct advantage in the production of recombinant protein. For example, if the recombinant protein has a harmful effect on the bacterium, then its synthesis must be carefully monitored to prevent accumulation of toxic levels; this can be achieved by judicious use of the regulatory chemical to control expression of the cloned gene. Even if the recombinant protein has no harmful effects on the host cell, regulation of the cloned gene is still desirable, as a continuously high level of transcription may affect the ability of the recombinant plasmid to replicate, leading to its eventual loss from the culture.

(b) Examples of promoters used in expression vectors   Several *E. coli* promoters combine the desired features of strength and ease of regulation. Those most frequently used in expression vectors are as follows.

1. The *lac* **promoter** (Figure 12.7(a)), which is the sequence that controls transcription of the *lacZ* gene coding for β-galactosidase

Figure 12.7 Four promoters often
used in expression vectors. The *lac*
and *trp* promoters are shown
upstream for the genes that they normally control in *E. coli.*

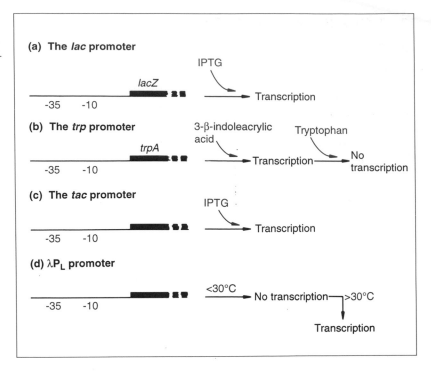

Figure 12.7 Four promoters often used in expression vectors. The *lac* and *trp* promoters are shown upstream for the genes that they normally control in *E. coli.*

(and also the *lacZ'* gene fragment carried by the pUC and M13mp vectors – pp. 112 and 116). The *lac* promoter is induced by IPTG (p. 98), so addition of this chemical into the growth medium switches on transcription of a gene inserted downstream of the *lac* promoter carried by an expression vector.

2. The **trp promoter** (Figure 12.7(b)), normally upstream of the cluster of genes coding for several of the enzymes involved in biosynthesis of the amino acid tryptophan. The *trp* promoter is repressed by tryptophan, but is more easily induced by 3-β-indoleacrylic acid.

3. The **tac promoter** (Figure 12.7(c)), a hybrid between the *trp* and *lac* promoters that is in fact stronger than either, but still induced by IPTG.

4. The **λP$_L$ promoter** (Figure 12.7(d)), one of the promoters responsible for transcription of the λ DNA molecule. λP$_L$ is a very strong promoter that is recognized by the *E. coli* RNA polymerase, which is subverted by λ into transcribing the bacteriophage DNA. The promoter is repressed by the product of the λ*cI* gene. Expression vectors that carry the λP$_L$ promoter are used with a mutant *E. coli* host that synthesizes a temperature-sensitive form of the *cI* protein (p. 45). At a low temperature (less than 30°C) this mutant *cI* protein is able to repress the λP$_L$ promoter; at higher temperatures the protein is inactivated resulting in transcription of the cloned gene.

12.1.2 Cassettes and gene fusions

An efficient expression vector requires not only a strong, regulatable promoter, but also an *E. coli* ribosome binding sequence and a terminator. In most vectors these expression signals form a **cassette**, so called because the foreign gene is inserted into a unique restriction site present in the middle of the expression signal cluster (Figure 12.8). Ligation of the foreign gene into the cassette therefore places it in the ideal position relative to the expression signals.

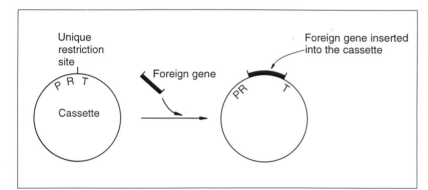

Figure 12.8 A typical cassette vector and the way it is used. P, promoter; R, ribosome binding site; T, terminator.

   With some cassette vectors the cloning site is not immediately adjacent to the ribosome binding sequence but instead is preceded by a segment from the beginning of an *E. coli* gene (Figure 12.9). Insertion of the foreign gene into this restriction site must be performed in such a way as to fuse the two reading frames, producing a hybrid gene that starts with the *E. coli* segment and progresses without a break into the codons of the foreign gene. The product of gene expression is therefore a hybrid protein, consisting of the short peptide coded by the *E. coli* reading frame fused to the amino-terminus of the foreign protein. This fusion system has four advantages:

1. Efficient translation of the mRNA produced from the cloned gene depends not only on the presence of a ribosome binding site, but is also affected by the nucleotide sequence at the start of the coding region. This is probably because secondary structures resulting from intrastrand base pairs could interfere with attachment of the ribosome to its binding site (Figure 12.10). This possibility is avoided if the pertinent region is made up entirely of natural *E. coli* sequences.
2. The presence of the bacterial peptide at the start of the fusion protein may stabilize the molecule and prevent it from being degraded by the host cell. In contrast, foreign proteins that lack a bacterial segment are often destroyed by the host.

Figure 12.9 The construction of a hybrid gene and the synthesis of a fusion protein. P, promoter; R, ribosome binding site; T, terminator.

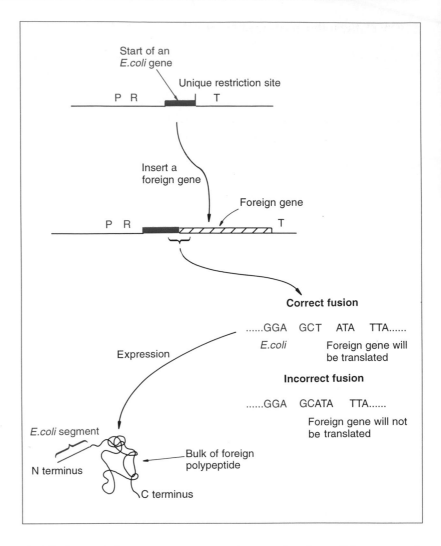

3. The bacterial segment may constitute a signal peptide, responsible for directing the *E. coli* protein to its correct position in the cell. If the signal peptide is derived from a protein that is exported by the cell (e.g. the products of the *ompA* or *malE*

Figure 12.10 A problem caused by secondary structure at the start of an mRNA.

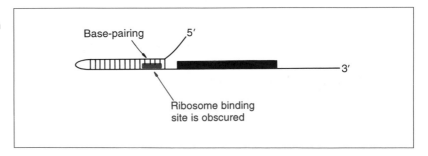

genes) then the recombinant protein may itself be exported, either into the culture medium or into the periplasmic space between the inner and outer cell membranes. Export is desirable as it simplifies the problem of purification of the recombinant protein from the culture.

4. The bacterial segment may also aid purification by enabling the fusion protein to be recovered by **affinity chromatography**. For example, fusions involving the *E. coli* glutathione-*S*-transferase protein can be purified by adsorption onto agarose beads carrying bound glutathione (Figure 12.11).

The disadvantage with a fusion system is that the presence of the *E. coli* segment may alter the properties of the recombinant protein. Methods for removing the bacterial segment are therefore needed. Usually this is achieved by treating the fusion protein with a chemical or enzyme that cleaves the polypeptide chain at or near to the junction between the two components. For example, if a methionine is present at the junction then the fusion protein can be cleaved with cyanogen bromide, which cuts

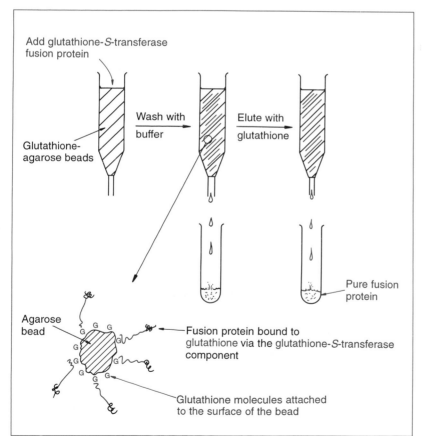

Figure 12.11 The use of affinity chromatography to purify a glutathione-*S*-transferase fusion protein.

Add glutathione-*S*-transferase fusion protein

Glutathione-agarose beads

Wash with buffer

Elute with glutathione

Pure fusion protein

Agarose bead

Fusion protein bound to glutathione via the glutathione-*S*-transferase component

Glutathione molecules attached to the surface of the bead

Figure 12.12 One method for the recovery of the foreign polypeptide from a fusion protein. The methionine residue at the fusion junction must be the only one present in the entire polypeptide: if others are present then cyanogen bromide will cleave the fusion protein into more than two fragments.

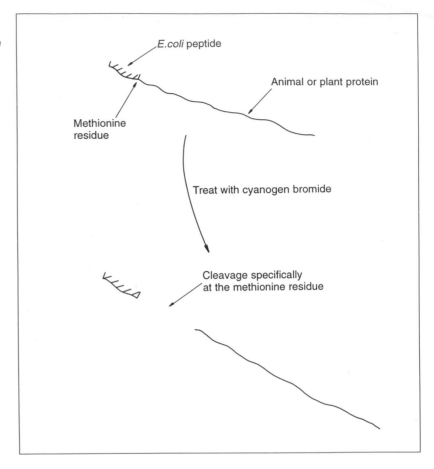

E.coli peptide

Animal or plant protein

Methionine residue

Treat with cyanogen bromide

Cleavage specifically at the methionine residue

polypeptides specifically at methionine residues (Figure 12.12). Alternatively, enzymes such as thrombin (which cleaves adjacent to arginine residues) or Factor Xa (which cuts after the arginine of Gly–Arg) can be used. The important consideration is that recognition sequences for the cleavage agent must not occur within the recombinant protein.

12.2 GENERAL PROBLEMS WITH THE PRODUCTION OF RECOMBINANT PROTEIN IN *E. COLI*

Despite the development of sophisticated expression vectors, there are still numerous difficulties associated with the production of protein from foreign genes cloned in *E. coli*. These problems can be grouped into two categories: those that are due to the sequence of the foreign gene, and those that are due to the limitations of *E. coli* as a host for recombinant protein synthesis.

12.2.1 Problems resulting from the sequence of the foreign gene

There are three ways in which the nucleotide sequence might prevent efficient expression of a foreign gene cloned in *E. coli*:

1. The foreign gene might contain introns. This would be a major problem as *E. coli* genes do not contain introns and the bacterium therefore does not possess the necessary machinery for removing introns from transcripts (Figure 12.13(a)).
2. The foreign gene might contain sequences that act as termination signals in *E. coli* (Figure 12.13(b)). These sequences are perfectly innocuous in the normal host cell but in the bacterium result in premature termination and a loss of gene expression.
3. The codon usage of the gene may not be ideal for translation in *E. coli*. Although virtually all organisms use the same genetic code, each organism has a bias towards preferred codons. This bias reflects the efficiency with which the tRNA molecules in the organism are able to recognize the different codons. If a cloned gene contains a high proportion of unfavoured codons, then the host cell's tRNAs may encounter difficulties in translating the gene, reducing the amount of protein that is synthesized (Figure 12.13(c)).

These problems can usually be solved, though the necessary manipulations may be time-consuming and costly (an important consideration in an industrial project). If the gene contains introns then its cDNA, prepared from the mRNA (p. 167) and so lacking introns, might be obtainable as an alternative. Oligonucleotide-directed mutagenesis could then be used to change the sequences of possible terminators and to replace unfavoured codons with those preferred by *E. coli*. An alternative with genes that are less than 2 kb in length is to make an artificial version by chemical synthesis (p. 177), using the amino acid sequence of the protein as the blueprint, and ensuring that preferred *E. coli* codons are used and that no termination sequences are present.

12.2.2 Problems caused by *E. coli*

Some of the difficulties encountered when using *E. coli* as the host for recombinant protein synthesis stem from inherent properties of the bacterium. For example:

1. *E. coli* might not process the recombinant protein correctly. The proteins of most organisms are processed after translation, by chemical modification of amino acids within the polypeptide. Often these processing events are essential for the correct biological activity of the protein. Unfortunately, the proteins of

Figure 12.13 Three of the problems that may be encountered when foreign genes are expressed in *E. coli.* (a) Introns are not removed in *E. coli.* (b) Premature termination of transcription. (c) A problem with codon bias. P, promoter; R, ribosome binding site; T, terminator.

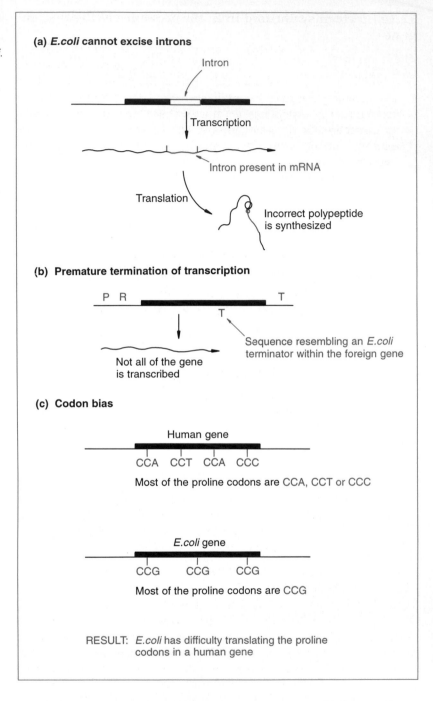

(a) *E.coli* cannot excise introns

Intron

Transcription

Intron present in mRNA

Translation

Incorrect polypeptide is synthesized

(b) Premature termination of transcription

P R   T

T

Not all of the gene is transcribed

Sequence resembling an *E.coli* terminator within the foreign gene

(c) Codon bias

Human gene

CCA   CCT   CCA   CCC

Most of the proline codons are CCA, CCT or CCC

E.coli gene

CCG   CCG   CCG

Most of the proline codons are CCG

RESULT: *E.coli* has difficulty translating the proline codons in a human gene

bacteria and higher organisms are not processed identically. In particular, some animal proteins are glycosylated, meaning that they have sugar groups attached to them after translation. Glycosylation is extremely uncommon in bacteria and recombinant proteins synthesized in *E. coli* are never glycosylated correctly.

2. *E. coli* might not fold the recombinant protein correctly. If the protein does not take up its correctly folded, tertiary structure then usually it is insoluble and forms an **inclusion body** within the bacterium (Figure 12.14). Recovery of the protein from the inclusion body is not a problem, but converting the protein into its correctly folded form is difficult or impossible in the test-tube. Under these circumstances the protein is, of course, inactive.

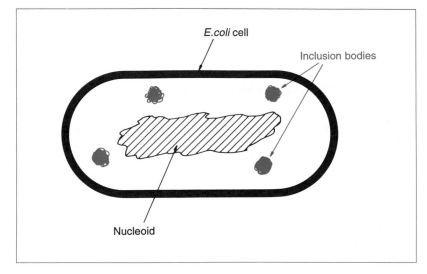

E.coli cell

Inclusion bodies

Nucleoid

Figure 12.14 Inclusion bodies.

3. *E. coli* might degrade the recombinant protein. It has already been mentioned that a fusion protein may be degraded less rapidly in *E. coli* than an unaltered recombinant protein. Exactly how *E. coli* can recognize the foreign protein, and thereby subject it to preferential turnover, is not known.

These problems are less easy to solve than the sequence problems described in the previous section. Degradation of recombinant proteins can be reduced by using as the host a mutant *E. coli* strain that is deficient in one or more of the proteases responsible for protein degradation. Correct folding of recombinant proteins can also be promoted by choosing a special host strain, in this case one that over-synthesizes the chaperone proteins thought to be responsible for protein folding in the cell. But the main problem is the absence of glycosylation. So far this has proved insurmountable, limiting *E. coli* to the synthesis of animal proteins that do not need to be processed in this way.

12.3 PRODUCTION OF RECOMBINANT PROTEIN BY EUKARYOTIC CELLS

The problems associated with obtaining high yields of active recombinant proteins from genes cloned in *E. coli* have led to the development of expression systems for higher organisms. The argument is that a microbial eukaryote, such as a yeast or filamentous fungus, is more closely related to an animal, and so may be able to deal with recombinant protein synthesis more efficiently than *E. coli*. Yeasts and fungi can be grown just as easily as bacteria in continuous culture, and may express a cloned gene from a higher organism, and process the resulting protein, in a manner more akin to that occurring in the higher organism itself.

12.3.1 Recombinant protein from yeast and filamentous fungi

To a certain extent these hopes have been realized and microbial eukaryotes are now being used for the routine production of several animal proteins. Expression vectors are still required as it turns out that the promoters and other expression signals for animal genes do not in general work efficiently in these lower eukaryotes. The vectors themselves are based on those described in Chapter 7.

The yeast *Saccharomyces cerevisiae* is currently the most popular microbial eukaryote for recombinant protein production. Cloned genes are often placed under the control of the *GAL* promoter (Figure 12.15(a)), which is normally upstream of the gene coding for galactose epimerase, an enzyme involved in the metabolism of galactose. The *GAL* promoter is induced by galactose, providing a straightforward system for regulating expression of a cloned foreign gene.

Yields of recombinant protein are relatively high, but *S. cerevisiae* is unable to glycosylate animal proteins correctly and lacks an efficient system for secreting proteins into the growth medium. In the absence of secretion, recombinant proteins are retained in the cell and are consequently less easy to purify. Codon bias (p. 265) can also be a problem. Despite these drawbacks, *S. cerevisiae* remains the most frequently used microbial eukaryote for recombinant protein synthesis, partly because it is accepted as a safe organism for production of proteins for use in medicines or in foods, and partly because of the wealth of knowledge built up over the years regarding the biochemistry and genetics of *S. cerevisiae*, which means that it is relatively easy to devise strategies for minimizing the difficulties that arise.

Although *S. cerevisiae* retains the loyalty of many molecular biol-

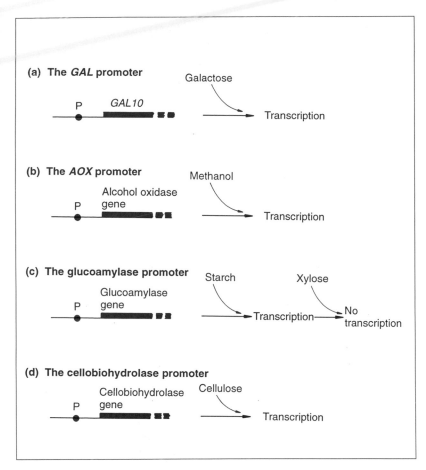

ogists, there are other microbial eukaryotes that might be equally if not more effective in recombinant protein synthesis. In particular, *Pichia pastoris*, a second species of yeast, holds a great deal of promise as it is able to synthesize large amounts of recombinant protein (up to 30% of the total cell protein) and its glycosylation abilities are very similar to those of animal cells. The sugar structures that it synthesizes are not precisely the same as the animal versions (Figure 12.16), but the differences are relatively trivial and would probably not have a significant effect on the activity of a recombinant protein. Importantly, the glycosylated proteins made by *P. pastoris* are unlikely to induce an antigenic reaction if injected into the bloodstream, a problem that is frequently encountered with the over-glycosylated proteins synthesized by *S. cerevisiae*. Expression vectors for *P. pastoris* make use of the alcohol oxidase (*AOX*) promoter (Figure 12.15(b)), which is induced by methanol.

The two most popular filamentous fungi are *Aspergillus nidulans* and *Trichoderma reesei*. The advantages of these organisms are their good glycosylation properties and their ability to secrete proteins

Figure 12.16 Comparison between a typical glycosylation structure found on an animal protein and the structures synthesized by *P. pastoris* and *S. cerevisiae*. Adapted from Cregg *et al.* (1993) *Biotechnology*, **11**, 905–10.

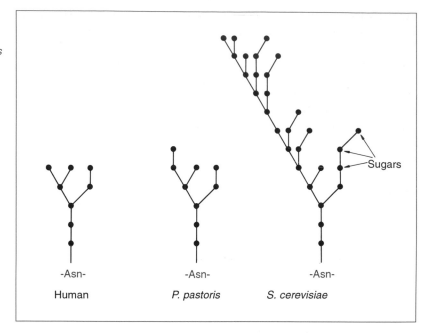

into the growth medium. The latter is a particularly strong feature of the wood-rot fungus *T. reesei*, which in its natural habitat secretes cellulolytic enzymes that degrade the wood that it lives on. The secretion characteristics mean that these fungi are able to produce recombinant proteins in a form that aids purification. Expression vectors for *A. nidulans* usually carry the glucoamylase promoter (Figure 12.15(c)), induced by starch and repressed by xylose; those for *T. reesei* make use of the cellobiohydrolase promoter (Figure 12.15(d)), which is induced by cellulose.

12.3.2 Using animal cells for recombinant protein production

The difficulties inherent in synthesis of a fully active animal protein in a microbial host have prompted biotechnologists to explore the possibility of using animal cells for recombinant protein synthesis. For proteins with complex and essential glycosylation structures an animal cell might be the only type of host within which the active protein can be synthesized.

Culture systems for animal cells have been around since the early 1960s but only recently have methods for large-sale continuous culture become available. A problem with some animal cell lines is that they require a solid surface on which to grow, adding complications to the design of the culture vessels. One solution is to fill the inside of the vessel with plates, providing a large surface area, but this has the disadvantage that complete and continuous mixing of the medium within the vessel becomes very difficult. A second possibility is to use a standard vessel but to provide the cells with small inert particles (e.g. cellulose beads) on which to grow. Rates of growth and maximum cell densities are much less for animal cells compared with microorganisms, limiting the yield of recombinant protein, but this can be tolerated if it is the only way of obtaining the active protein.

Of course, gene cloning may not be necessary in order to obtain an animal protein from an animal cell culture. Nevertheless, expression vectors and cloned genes are still used to maximize yields, by placing the gene under control of a promoter that is stronger than the one it is normally attached to. Two promoters that have been used in mammalian cells are the heat-shock promoter of the human *hsp-70* gene, which is induced at temperatures above 40°C, and the mouse metallothionein gene promoter, which is switched on by addition of zinc salts to the culture medium (Figure 12.17).

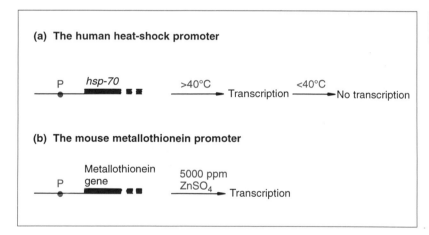

(a) The human heat-shock promoter

P   *hsp-70*   >40°C   <40°C
                → Transcription → No transcription

(b) The mouse metallothionein promoter

P   Metallothionein gene   5000 ppm ZnSO$_4$
                → Transcription

Figure 12.17 Two promoters that have been used in expression vectors for mammalian cells.

Insect cells provide a potentially important alternative to mammalian cells for animal protein production. Insect cells do not behave in culture any differently to mammalian cells but they have the great advantage that, thanks to a natural expression system, they can provide high yields of recombinant protein. The expression system is based on the **baculoviruses**, a group of viruses that are common in insects but apparently do not infect vertebrates. The baculovirus genome includes the polyhedrin gene, whose normal product accumulates in the insect cell as large nuclear inclusion bodies towards the end of the infection cycle (Figure 12.18). The product of this single gene frequently makes up over 50% of the total cell protein. Similar levels of protein production also occur if the normal gene is replaced by a foreign one. The big question is

Figure 12.18 Crystalline inclusion bodies in the nuclei of insect cells infected with a baculovirus.

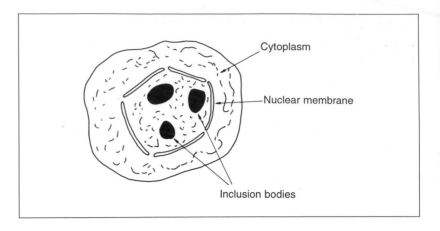

Cytoplasm

Nuclear membrane

Inclusion bodies

whether insect cells process proteins in the same way as mammalian cells: it appears that in many cases they do, but not always. Further research will reveal exactly how useful the baculovirus expression system is going to be.

FURTHER READING

Primrose, S. B. (1991) *Molecular Biotechnology*, 2nd edn, Blackwell Scientific Publications, Oxford – gives details of all aspects of biotechnology, including recombinant protein production and large-scale culture systems.

DeBoer, H. A. *et al.* (1983) The *tac* promoter: a functional hybrid derived from the *trp* and *lac* promoters. *Proceedings of the National Academy of Sciences, USA*, **80**, 21–5.

Remaut, E. *et al.* (1981) Plasmid vectors for high-efficiency expression controlled by the P_L promoter of coliphage. *Gene*, **15**, 81–93 – construction of an expression vector.

Lee, N., Cozzikorto, J., Wainwright, N. and Testa, D. (1984) Cloning with tandem gene systems for high level gene expression. *Nucleic Acids Research*, **12**, 6797–812 – fusion proteins.

Smith, D. B. and Johnson, K. S. (1988) Single-step purification of polypeptides expressed in *Escherichia coli* as fusions with glutathione *S*-transferase. *Gene*, **67**, 31–40.

Robinson, M., Lilley, R., Little, S., Emtage, J. S., Yarranton, G., Stephens, P., Millican, A., Eaton, M. and Humphreys, G. (1984) Codon usage can affect efficiency of translation of genes in *Escherichia coli*. *Nucleic Acids Research*, **12**, 6663–71.

Meerman, H. J. and Georgiou, G. (1994) Construction and characterization of a set of *E. coli* strains deficient in all known loci affecting the proteolytic stability of secreted recombinant proteins. *Biotechnology*, **12**, 1107–10 – countering the problems caused by degradation of recombinant proteins in *E. coli*.

Hitzeman, R. A., Hagie, F. E., Levine, H. L., Goeddel, D. V., Ammerer, G. and Hall, B. D. (1981) Expression of human gene for interferon in yeast. *Nature*, **293**, 717–22 – the first use of *S. cerevisiae* in the production of a recombinant human protein.

Cregg, J. M., Vedvick, T. S. and Raschke, W. C. (1993) Recent advances in the expression of foreign genes in *Pichia pastoris*. *Biotechnology*, **11**, 905–10.

Saunders, G. *et al.* (1989) Heterologous gene expression in filamentous fungi. *Trends in Biotechnology*, **7**, 283–7.

Cameron, I. R. *et al.* (1989) Insect cell culture technology in baculovirus expression systems. *Trends in Biotechnology*, **7**, 66–70.

13 Gene cloning in medicine

The aim of the final two chapters is to illustrate the impact that gene cloning is having in research and biotechnology, stimulating advances and breakthroughs that could never be achieved without the ability to purify and manipulate individual genes. This is a topic about which an entire book could be written, but the endeavour would be a waste of time because the book would be out of date within a few weeks, so rapid is the current progress. We must limit ourselves to just two areas in which the new biology is having a major influence: medicine in this chapter and agriculture in the next.

Medicine has been and will continue to be a major beneficiary of the gene cloning revolution. Later in this chapter we will see how recombinant DNA techniques are being used to identify genes responsible for inherited diseases and to devise new therapies for these disorders. First we will continue the theme developed in the last chapter and examine the ways in which cloned genes are being used in the production of recombinant pharmaceuticals.

13.1 PRODUCTION OF RECOMBINANT PHARMACEUTICALS

A number of human disorders can be traced to the absence or malfunction of a protein normally synthesized in the body. Most of these disorders can be treated by supplying the patient with the correct version of the protein, but for this to be possible the relevant protein must be available in relatively large amounts. If the defect can be corrected only by administering the human protein, then obtaining sufficient quantities will be a major problem unless donated blood can be used as the source. Animal proteins are therefore used whenever these are effective, but there are not many disorders that can be treated with animal proteins, and there is always the possibility of side effects such as an allergenic response.

We learnt in Chapter 12 that gene cloning can be used to obtain large amounts of recombinant human proteins. How are these techniques being applied to the production of proteins for use as pharmaceuticals?

13.1.1 Recombinant insulin

Insulin, synthesized by the β-cells of the Islets of Langerhans in the pancreas, controls the level of glucose in the blood. An insulin deficiency manifests itself as diabetes mellitus, a complex of symptoms which may lead to death if untreated. Fortunately, many forms of diabetes can be alleviated by a continuing programme of insulin injections, thereby supplementing the limited amount of hormone synthesized by the patient's pancreas. The insulin used in this treatment has traditionally been obtained from the pancreas of pigs and cows slaughtered for meat production. Although animal insulin is generally satisfactory, problems may arise in its use to treat human diabetes. One is that the slight differences between the animal and the human proteins may lead to side effects in some patients. The second is that the purification procedures are difficult and potentially dangerous contaminants cannot always be completely removed.

Insulin displays two features that facilitate its production by recombinant DNA techniques. The first is that the human protein is not modified after translation by the addition of sugar molecules (p. 266); recombinant insulin synthesized by a bacterium should therefore be active. The second advantage concerns the size of the molecule. Insulin is a relatively small protein, comprising two polypeptides, one of 21 amino acids (the A chain) and one of 30 amino acids (the B chain, Figure 13.1). In humans these are synthesized as a precursor called preproinsulin which contains the A and B segments linked by a third chain (C) and preceded by a leader sequence. The leader sequence is removed after translation and the C chain excised, leaving the A and B polypeptides linked to each other by two disulphide bonds.

Several strategies have been used to obtain recombinant insulin. One of the first projects, involving synthesis of artificial genes for the A and B chains followed by production of fusion proteins in *E. coli*, illustrates a number of the general techniques used in recombinant protein production.

(a) Synthesis and expression of artificial insulin genes   In the late 1970s the idea of making an artificial gene was extremely innovative. Oligonucleotide synthesis was in its infancy at that time, and the available methods for making artificial DNA molecules were much more cumbersome than the present-day automated techniques. Nevertheless, genes coding for the A and B chains of insulin were synthesized as early as 1978.

Figure 13.1 The structure of the
insulin molecule and a summary of
its synthesis by processing from preproinsulin.

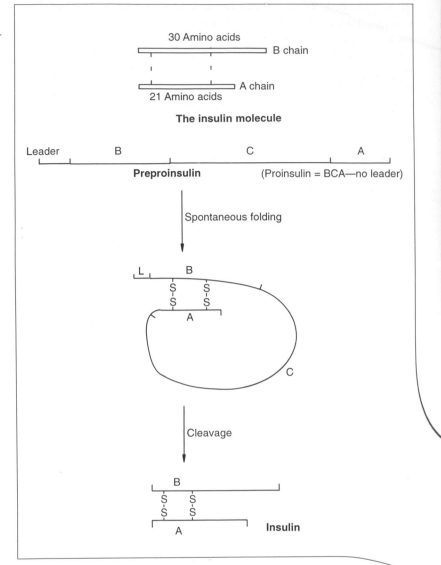

The procedure was to synthesize trinucleotides repr...
the possible codons and then join these together in th...
tated by the amino acid sequences of the A and B chai...
cial genes would not necessarily have the sam...
sequences as the real gene segments coding for the ...
but would still specify the correct polypeptides. T...
plasmids were constructed, one carrying the artif...
A chain and one the gene for the B chain. In each...
gene was ligated to a *lacZ'* reading frame preser...
vector (Figure 13.2(a)). The insulin genes were...
control of the strong *lac* promoter (p. 259), a...

fusion proteins, consisting of the first few amino acids of β-galactosidase followed by the A or B polypeptides (Figure 13.2(b)). Each gene was designed so that its β-galactosidase and insulin segments were separated by a methionine residue, so that the insulin polypeptides could be cleaved from the β-galactosidase segments by treatment with cyanogen bromide (p. 263). The purified A and B chains were then attached to each other by disulphide bond formation in the test-tube.

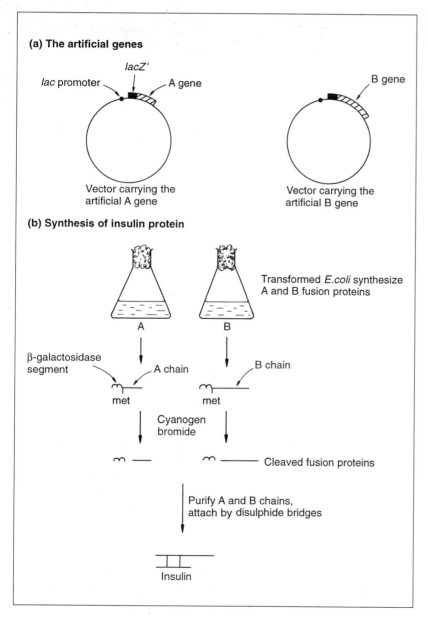

Figure 13.2 The synthesis of recombinant insulin from artificial A and B chain genes.

(a) The artificial genes

lacZ'

lac promoter

A gene

B gene

Vector carrying the artificial A gene

Vector carrying the artificial B gene

(b) Synthesis of insulin protein

Transformed *E.coli* synthesize A and B fusion proteins

A

B

β-galactosidase segment

A chain

B chain

met

met

Cyanogen bromide

Cleaved fusion proteins

Purify A and B chains, attach by disulphide bridges

Insulin

The final step, involving disulphide bond formation, is actually rather inefficient. A subsequent improvement has been to synthesize not the individual A and B genes, but the entire proinsulin reading frame, specifying B chain–C chain–A chain (Figure 13.1). Although this is a more daunting proposition in terms of DNA synthesis, the prohormone has the big advantage of folding spontaneously into the correct, disulphide-bonded structure. The C chain segment can then be excised relatively easily by proteolytic cleavage.

13.1.2 Synthesis of human growth hormones in *E. coli*

At about the same time that recombinant insulin was first being made in *E. coli*, other researchers were working on similar projects with the human growth hormones somatostatin and somatotropin. These two proteins act in conjunction to control growth processes in the human body, their malfunction leading to painful and disabling disorders such as agromegaly (uncontrolled bone growth) and dwarfism.

Somatostatin was the first human protein to be synthesized in *E. coli*. Being a very short protein, only 14 amino acids in length, it was ideally suited for artificial gene synthesis. The strategy used was the same as described for recombinant insulin, involving insertion of the artificial gene into a *lacZ'* vector (Figure 13.3), syn-

Figure 13.3 Production of recombinant somatostatin.

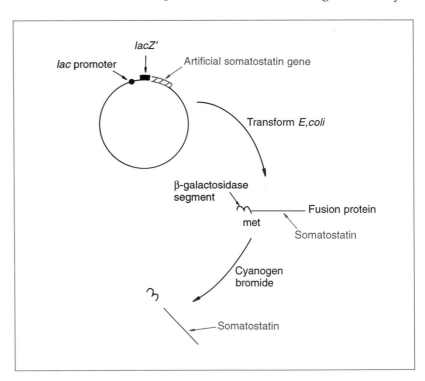

thesis of a fusion protein, and cleavage with cyanogen bromide.

Somatotropin presented a more difficult problem. This protein is 191 amino acids in length, equivalent to almost 600 bp, a difficult prospect for today's DNA synthesis capabilities, let alone those of the late 1970s. In fact a combination of artificial gene synthesis and cDNA cloning was used to obtain a somatotropin-producing *E. coli* strain. mRNA was obtained from the pituitary, the gland that produces somatotropin in the human body, and a cDNA library prepared. The somatotropin cDNA turned out to have a unique site for the restriction endonuclease *Hae*III, which cuts the gene into two segments (Figure 13.4(a)). The longer segment, consisting of

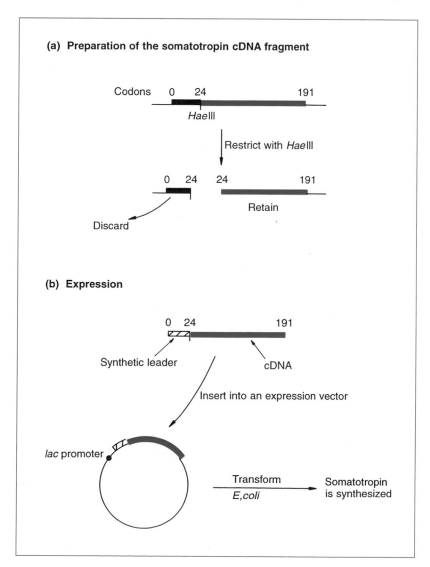

Figure 13.4 Production of recombinant somatotropin.

codons 24 to 191, was retained for use in construction of the recombinant plasmid. The smaller segment was replaced by an artificial DNA molecule that reproduced the start of the somatotropin gene and provided the correct signals for translation in *E. coli* (Figure 13.4(b)). The modified gene was then ligated into an expression vector carrying the *lac* promoter.

13.1.3 Recombinant Factor VIII

Although a number of important pharmaceutical compounds have been obtained from genes cloned in *E. coli*, the general problems associated with using bacteria to synthesize foreign proteins (p. 264) have led in many cases to these organisms being replaced by eukaryotes. An example of a recombinant pharmaceutical produced in eukaryotic cells is human Factor VIII, a protein which plays a central role in blood clotting. The commonest form of haemophilia in humans results from an inability to synthesize Factor VIII, leading to a breakdown in the blood clotting pathway and the well-known symptoms associated with the disease. At present the only way to treat haemophilia is by injection of purified Factor VIII protein, obtained from human blood provided by donors. Purification of Factor VIII is a complex procedure and the treatment is very expensive, with estimates of over $10 000 per annum for each patient. More critically, the purification is beset with difficulties, in particular in removing virus particles that may be present in the blood. Hepatitis and AIDS can and have been passed on to haemophiliacs via Factor VIII injections. Recombinant Factor VIII, free from contamination problems, would be a significant achievement for biotechnology.

The Factor VIII gene is very large, over 186 kb in length, and is split into 26 exons and 25 introns (Figure 13.5(a)). The mRNA codes for a large polypeptide (2351 amino acids) which undergoes a complex series of post-translational processing events, eventually resulting in a dimeric protein consisting of a large subunit, derived from the upstream region of the initial polypeptide, and a small subunit from the downstream segment (Figure 13.5(b)). The two subunits contain a total of 17 disulphide bonds and a number of glycosylated sites. As might be anticipated for such a large and complex protein, it has not been possible to synthesize an active version in *E. coli*.

Most attempts to obtain recombinant Factor VIII have therefore involved mammalian cells. In the first experiments to be carried out the entire cDNA was cloned in hamster cells, but yields of protein were disappointingly low. This was probably because the post-translational events, although carried out correctly in hamster cells, do not convert all of the initial product into an active form, limiting the overall yield. As an alternative, two separate

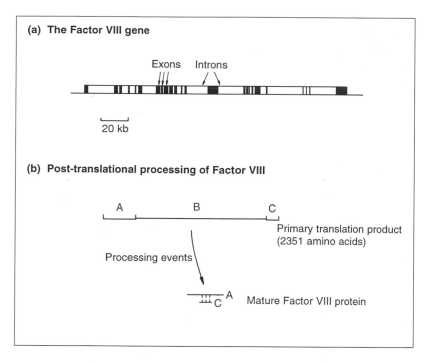

Figure 13.5 The Factor VIII gene and its translation product.

fragments from the cDNA were used, one fragment coding for the large subunit polypeptide, the second for the small subunit. Each cDNA fragment was ligated into an expression vector, downstream of the Ag promoter (a hybrid between the chicken β-actin and rabbit β-globin sequences) and upstream of a polyadenylation signal from SV40 virus (Figure 13.6). The plasmid was introduced into a hamster cell line and recombinant protein obtained. The yields were over ten times greater than those from cells containing the complete cDNA, and the resulting Factor VIII protein was indistinguishable in terms of function from the native form.

13.1.4 Synthesis of other recombinant human proteins

The list of human proteins synthesized by recombinant technology continues to grow (Table 13.1). As well as proteins used to treat

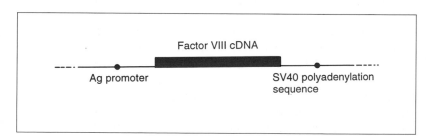

Figure 13.6 The expression signals used in production of recombinant Factor VIII. The promoter is an artificial hybrid of the chicken β-actin and rabbit β-globin sequences, and the polyadenylation signal (needed for correct processing of the mRNA before translation into protein) is obtained from SV40 virus.

Table 13.1 Some of the human proteins that have been synthesized from genes cloned in bacteria and/or eukaryotic cells

Protein	Used in the treatment of
Insulin	Diabetes
Somatostatin	Growth disorders
Somatrotropin	Growth disorders
Factor VIII	Haemophilia
Factor IX	Christmas disease
Interferon-α	Leukaemia and other cancers
Interferon-β	Cancers, AIDS
Interferon-γ	Cancers, rheumatoid arthritis
Interleukins	Cancers, immune disorders
Granulocyte colony stimulating factor	Cancers
Tumour necrosis factor	Cancers
Epidermal growth factor	Ulcers
Fibroblast growth factor	Ulcers
Erythropoietin	Anaemia
Tissue plasminogen activator	Heart attack
Superoxide dismutase	Free radical damage in kidney transplants
Lung surfactant protein	Respiratory distress
α1-antitrypsin	Emphysema
Serum albumin	Used as a plasma supplement
Relaxin	Used to aid childbirth

disorders by replacement or supplementation of the malfunctional versions, the list includes a number of growth factors (e.g. interferons and interleukins) with potential uses in cancer therapy. These proteins are synthesized in very limited amounts in the body, so recombinant technology is the only viable means of obtaining them in the quantities needed for clinical purposes. Other proteins, such as serum albumin, are more easily obtained, but are needed in such large quantities that production in microorganisms is still a more attractive option.

13.1.5 Recombinant vaccines

The final category of recombinant protein is slightly different from the examples given in Table 13.1. A vaccine is an antigenic preparation that, after injection into the bloodstream, stimulates the immune system to synthesize antibodies that protect the body against infection. The antigenic material present in a vaccine is normally an inactivated form of the infectious agent. For example, antiviral vaccines often consist of virus particles that have been attenuated by heating or a similar treatment. In the past, two problems have hindered the preparation of attenuated viral vaccines:

1. The inactivation process must be 100% efficient, as the presence in a vaccine of just one live virus particle could result in infection. This has been a problem with vaccines for the cattle disease, foot-and-mouth.
2. The large amounts of virus particles needed for vaccine production are usually obtained from tissue cultures. Unfortunately some viruses, notably hepatitis B virus, do not grow in tissue culture.

(a) Producing vaccines as recombinant proteins The use of gene cloning in this field centres on the discovery that virus-specific antibodies are sometimes synthesized in response not only to the whole virus particle, but also to isolated components of the virus. This is particularly true of purified preparations of the proteins present in the virus coat (Figure 13.7). If the genes coding for the antigenic proteins of a particular virus could be identified and inserted into an expression vector, then the methods described above for the synthesis of animal proteins could be employed in the production of recombinant proteins that might be used as vaccines. These vaccines would have the advantages that they would be free of intact virus particles and they could be obtained in large quantities.

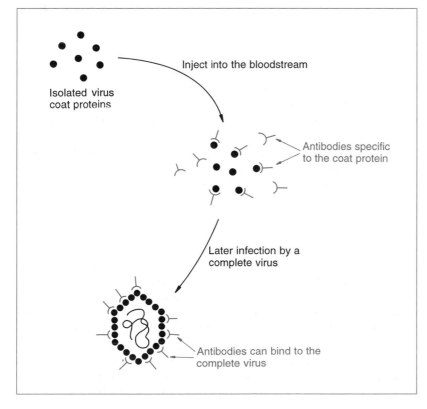

Figure 13.7 The principle behind the use of a preparation of isolated virus coat proteins as a vaccine.

Unfortunately this approach has not been entirely successful, mainly because it has turned out that recombinant coat proteins often lack the full antigenic properties of the intact virus. The one notable success has been with hepatitis B virus, whose coat protein (the 'major surface antigen') has been synthesized in *Saccharomyces cerevisiae*, using a vector based on the 2μm plasmid (p. 133). The protein was obtained in reasonably high quantities and when injected into monkeys provided protection against hepatitis B. This recombinant vaccine has been approved for use in humans, and the possibility of using hamster cells to produce an even better vaccine is being explored.

(b) Live recombinant vaccines The use of live vaccinia virus as a vaccine for smallpox dates back to 1796, when Jenner first realized that this virus, harmless to humans, could stimulate immunity against the much more dangerous smallpox virus. The term 'vaccine' comes from vaccinia; its use resulted in the worldwide eradication of smallpox in 1980.

A much more recent idea is that recombinant vaccinia viruses could be used as live vaccines against other diseases. If a gene coding for a virus coat protein, for example the hepatitis B major surface antigen, is ligated into the vaccinia genome, under control of a vaccinia promoter, then the gene will be expressed (Figure 13.8). After injection into the bloodstream, replication of the recombinant virus will result not only in new vaccinia particles, but also in significant quantities of the major surface antigen. Immunity against both smallpox and hepatitis B would result.

This remarkable technique has considerable potential. Already recombinant vaccinia viruses expressing a number of foreign genes have been constructed and shown to confer immunity against the relevant diseases in experimental animals (Table 13.2). The possibility of broad-spectrum vaccines is raised by the demonstration that a single recombinant vaccinia, expressing the genes for influenza virus haemagglutinin, hepatitis B major surface antigen and herpes simplex virus glycoprotein, confers immunity against each disease in monkeys. The important question that must now be answered is whether we know enough about viral biology to be sure that release of recombinant vaccinia viruses into the biosphere via vaccination programmes can be allowed.

13.2 IDENTIFICATION OF GENES RESPONSIBLE FOR HUMAN DISEASES

A second major area of medical research in which gene cloning is having an impact is in the identification and isolation of genes responsible for human diseases. A genetic or inherited disease is

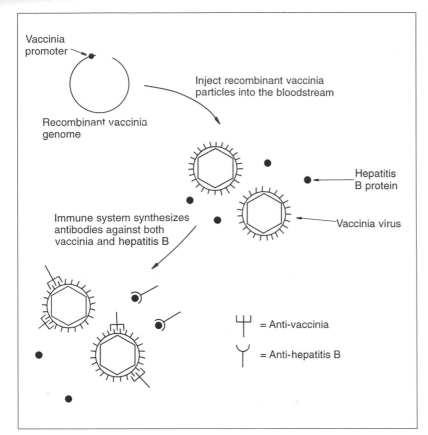

Figure 13.8 The rationale behind the potential use of a recombinant vaccinia virus.

one that is caused by a defect in a specific gene (Table 13.3), individuals carrying the defective gene being predisposed towards developing the disease at some stage of their lives. With some inherited diseases, such as haemophilia, the gene is present on the X chromosome, so all males carrying the gene express the disease

Table 13.2   Some of the foreign genes expressed by recombinant vaccinia viruses

Gene
Plasmodium falciparum surface antigen
Influenza virus coat proteins
Rabies virus G protein
Hepatitis B major surface antigen
Herpes simplex glycoproteins
HIV envelope proteins
Vesicular stomatitis coat proteins
Sindbis virus proteins

Table 13.3   Some of the commonest genetic diseases in the UK

Disease	Symptoms	Frequency (births/year)
Inherited breast cancer	Cancer	1 in 300 females
Cystic fibrosis	Lung disease	1 in 2000
Huntington's chorea	Neurodegeneration	1 in 2000
Duchenne muscular dystrophy	Progressive muscle weakness	1 in 3000 males
Haemophilia A	Blood disorder	1 in 4000 males
Sickle cell anaemia	Blood disorder	1 in 10 000
Phenylketonuria	Mental retardation	1 in 12 000
β-Thalassaemia	Blood disorder	1 in 20 000
Retinoblastoma	Cancer of the eye	1 in 20 000
Haemophilia B	Blood disorder	1 in 25 000 males
Tay–Sachs disease	Blindness, loss of motor control	1 in 200 000
McArdle disease	Progressive muscle weakness	1 in 500 000

state; females with one defective gene and one correct gene are healthy but can pass the disease to their male offspring. Genes for other diseases are present on autosomes and in most cases are recessive, so both chromosomes of the pair must carry a defective version for the disease to occur; a few diseases, including Huntington's chorea, are autosomal dominant so a single copy of the defective gene is enough to lead to the disease state. With some genetic diseases, the symptoms manifest themselves early in life, with others the disease may not be expressed until the individual is middle-aged or elderly. Cystic fibrosis is an example of the former, neurodegenerative diseases such as Alzheimer's and Huntington's are examples of the latter. With a number of diseases that appear to have a genetic component, cancers in particular, the overall syndrome is complex with the disease remaining dormant until triggered by some metabolic or environmental stimulus. If predisposition to these diseases can be diagnosed then the risk factor can be reduced by careful management of the patient's lifestyle to minimize the chances of the disease being triggered.

Genetic diseases have always been present in the human population but their importance has increased in recent decades. This is because vaccination programmes, antibiotics and improved sanitation have reduced the prevalence of infectious diseases such as smallpox, tuberculosis and cholera, which were major killers of the early twentieth century. The result is that a greater proportion of the population now dies from a disease that has a genetic component, especially the late-onset diseases which are now more common because of increased life expectancies. Medical research has been successful in controlling many infectious diseases; can it be equally successful with genetic disease?

There are a number of reasons why identifying the gene responsible for a genetic disease is important:

1. Gene identification may provide an indication of the biochemical basis to the disease, enabling therapies to be designed.
2. Identification of the mutation present in a defective gene can be used to devise a screening programme so that the mutant gene can be identified in individuals who are carriers or who have not yet developed the disease. Carriers can receive counselling regarding the chances of their children inheriting the disease. Early identification in individuals who have not yet developed the disease allows appropriate precautions to be taken to reduce the risk of the disease becoming expressed.
3. Identification of the gene is a prerequisite for gene therapy (p. 291).

During the last few years a number of genes responsible for inherited diseases have been identified. We will examine a typical example of one of these projects, concerning the breast cancer gene *BRCA1*.

13.2.1 The breast cancer gene *BRCA1*

More than one in ten women develop breast cancer with over half of these dying from the disease, some 25 000 deaths per year in the UK alone. Most women who contract breast cancer have no family history of the disease but in others a predisposition is inherited as an autosomal dominant trait. In these families, women who carry the defective gene have a 90% probability of developing the disease during their lifetimes. With such a mortality rate, both for the inherited and non-inherited forms, identification of the gene or genes responsible for breast cancer is clearly an urgent priority.

  (a) Mapping *BRCA1* The first task in isolation of any gene from the human genome is to determine which chromosome carries the gene and to map, as accurately as possible, the position of the gene on this chromosome. This is usually carried out by **linkage analysis**, which involves comparing the inheritance pattern for the target gene with the inheritance patterns for genetic loci whose map positions are already known. If two loci are inherited together then they must be very close on the same chromosome: if this is not the case then recombination events and the random segregation of chromosomes during meiosis will result in the loci displaying different inheritance patterns (Figure 13.9). Demonstration of linkage

Figure 13.9 Inheritance patterns for linked and unlinked genes. Three families are shown, circles representing females and squares representing males. (a) Two closely linked genes are almost always inherited together. (b) Two genes on different chromosomes display random segregation. (c) Two genes that are far apart on a single chromosome are often inherited together, but recombination may unlink them.

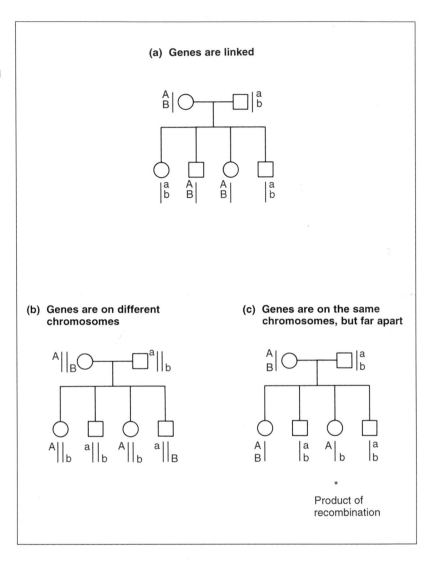

with one or more mapped genetic loci is therefore the key to understanding the chromosomal position of an unmapped gene.

Linkage analysis is carried out by examining DNA samples from individuals in families predisposed to the genetic disease. With breast cancer, the first breakthrough occurred in 1990 as a result of RFLP linkage analyses (p. 201) carried out by a group at the University of California at Berkeley. This study showed that in families with a high incidence of breast cancer a significant number of the women who suffered from the disease all possessed the same version of an RFLP called *D17S74*. This RFLP had previously been mapped to the long arm of chromosome 17 (Figure 13.10). *BRCA1* must therefore also be located on the long arm of chromosome 17.

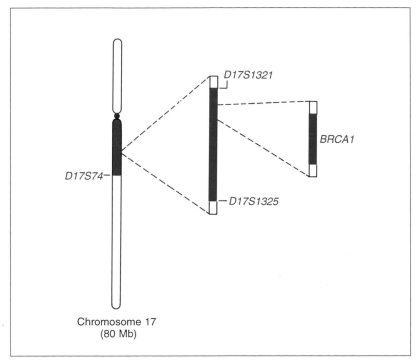

Figure 13.10 Mapping the breast cancer gene. Initially the gene was mapped to a 20-Mb segment of chromosome 17 (highlighted region in the left drawing). Additional mapping experiments narrowed this down to a 600-kb region flanked by two previously-mapped loci, *D17S1321* and *D17S1325* (middle drawing). After examination of expressed sequences, a strong candidate for *BRCA1* was eventually identified (right drawing).

This initial linkage result was extremely important as it indicated in which region of the human genome the breast cancer gene was to be found, but it was far from the end of the story. In fact over 1000 genes are thought to lie in this particular 20-Mb stretch of chromosome 17. The next objective was therefore to carry out more linkage studies to try to pinpoint *BRCA1* more accurately. This was achieved by first examining the region containing *BRCA1* for tandem repeat sequences, similar to those used in genetic fingerprinting (see p. 202 and Figure 9.14(a)), but with very short repeat units (e.g. ...CACACA...). These repeat sequences are useful in fine-detail linkage analysis as the number of repeats at a locus is highly variable, probably because occasional errors in DNA replication lead to additional units being inserted. This results in length variations that can be typed by PCR, using primers that flank the repeat region, with the size of the amplification product measured by gel electrophoresis (p. 236). Linkage is assessed by comparing the inheritance patterns of a locus of a particular length and the gene under investigation. This approach, using the repeat loci identified on chromosome 17, reduced the size of the *BRCA1*-containing region from 20 Mb down to just 600 kb (Figure 13.10).

(b) Identification of a candidate gene for *BRCA1*   Now that the relevant region had been narrowed down to just 600 kb it was possible to search the DNA segment for genes that might be

BRCA1. First, clones containing DNA from the segment were isolated from human libraries prepared with high-capacity cosmid, YAC, P1 and BAC vectors (pp. 130 and 139). The DNA fragments were then analysed in order to identify the genes that were present. One might imagine that this would simply be a case of sequencing the DNA and searching for open reading frames. Unfortunately this is not as easy as it sounds as many human genes are split into exons and introns, which complicates the identification of continuous open reading frames, even when computer search programmes are used. The sequence analysis was therefore supplemented by construction of a transcript map (only genes are transcribed!) for the *BRCA1* region. RNA was prepared from a variety of tissues, converted into cDNA (p. 167), the cDNAs sequenced, and those derived from the *BRCA1* region identified by comparison with the sequence of the genomic DNA. This combination of DNA sequence analysis and transcript mapping resulted in 65 expressed sequences being identified, probably representing somewhat fewer than 65 genes when account is taken of the likely fragmentation of long mRNAs during extraction, plus the possibility that one gene might give rise to a number of transcripts of different lengths.

Three approaches were used to identify a candidate for *BRCA1* from among these expressed sequences:

1. Hybridization probes specific for individual expressed sequences were used to examine RNA samples from different tissues. A candidate for *BRCA1* would be expected to hybridize to RNA prepared from breast tissue, and also to ovary tissue RNA, ovarian cancer frequently being associated with inherited breast cancer.
2. Southern hybridization analysis (p. 184) was carried out with DNA from different species. The rationale is that an important human gene will almost certainly have homologues in other mammals, and that this homologue, although having a slightly different sequence from the human version, will be detectable by hybridization with a suitable probe.
3. Genes thought to be possible candidates for *BRCA1* were sequenced from women suffering from breast cancer and these sequences compared with those obtained from non-susceptible individuals. The presence of mutations in the genes from cancer victims would be strong evidence that the *BRCA1* gene had been found.

These analyses resulted in identification of an approximately 100 kb gene, made up of 22 exons and coding for a 1863-amino-acid protein, that is a strong candidate for *BRCA1*. Transcripts of the gene are detectable in breast and ovary tissues, and homologues are present in mice, rats, rabbits, sheep and pigs, but not chickens. Most importantly, the genes from five susceptible families contain mutations (such as frameshifts and nonsense mutations) likely to lead to a non-functioning protein. Although circumstantial, the evidence in support of the candidate is sufficiently overwhelming for this gene to be identified as *BRCA1*.

(c) The next set of questions The identification of any gene responsible for an inherited disease is greeted by the scientific (and sometimes popular) press as a cause for celebration. In reality it is just the first step in a much longer series of research projects that must exploit this initial result in the search for ways of preventing, diagnosing and treating the disease. With *BRCA1*, the next set of experiments will try to build on the discovery that the protein coded by the gene is probably a transcription factor that regulates the expression of other genes. This is suggested by the presence of an amino acid sequence characteristic of a zinc finger (Figure 13.11), a structural motif possessed by several DNA-binding proteins. A key to future progress will be identification of the genes that are regulated by the *BRCA1* gene product.

...ECPICLELIKEPVSTKCDHIFCKFCMLKLLNQKKGPSQCPLCK...

Figure 13.11 The amino acid sequence of *BRCA1* between residues 23 and 65. The relative positions of the one histidine and seven cysteines that are highlighted are characteristic of a zinc finger domain.

Sometimes, identification of a disease-causing gene raises more questions than it answers. *BRCA1* is very much in this category. One problem was revealed by the mapping studies that placed *BRCA1* on chromosome 17, the results of these experiments suggesting that there might be more than one breast cancer gene. In fact, *BRCA2* has now been mapped to chromosome 13. Further difficulties emerged when *BRCA1* was examined in women with non-inherited breast cancer, the type that makes up over 90% of all cases. No convincing evidence could be obtained for mutations in *BRCA1* in these women, suggesting that the causations of the inherited and non-inherited diseases are different. Clearly a great deal of work still has to be done.

13.3 GENE THERAPY

The final application of cloning in medicine that we will consider is **gene therapy**. This is the name given to methods that aim to cure an inherited disease by providing the patient with a correct copy of the defective gene. Gene therapy has been successful with experi-

Art Center College of Design
Library
1700 Lida Street
Pasadena, Calif. 91103

mental animals and clinical trials with humans have been approved by the relevant regulatory agencies. First we will examine the techniques used in gene therapy, and then we will attempt to address the ethical issues.

13.3.1 Approaches to gene therapy

There are two basic approaches to gene therapy: germline therapy and somatic cell therapy. In germline therapy a fertilized egg is provided with a copy of the correct version of the relevant gene and reimplanted into the mother. If successful, the gene is present and expressed in all cells of the resulting individual. Germline therapy is usually carried out by microinjection of DNA into the isolated egg cell (p. 104), and theoretically could be used to treat any inherited disease.

Somatic cell therapy involves manipulation of ordinary cells, usually ones which can be removed from the organism, transfected, and then placed back in the body. The technique has most promise for inherited blood diseases (e.g. haemophilia, thalassaemia), with genes being introduced into stem cells from the bone marrow, which give rise to all the specialized cell types in the blood. The strategy is to prepare a bone extract containing several billion cells, transfect these with a **retrovirus**-based vector, and then reimplant the cells. Subsequent replication and differentiation of transfectants leads to the added gene being present in all the mature blood cells (Figure 13.12). The advantage of a retrovirus vector is that this type of vehicle has an extremely high transfection frequency, enabling a large proportion of the stem cells in a bone marrow extract to receive the new gene. Somatic cell therapy also has potential in the treatment of lung diseases such as cystic fibrosis, as DNA introduced into the respiratory tracts of rats via an inhaler is taken up by the epithelial cells in the lungs, although gene expression occurs for only a few weeks. Nevertheless, this gene delivery method is so simple that repeated use would not present a problem.

With those genetic diseases where the defect arises because the mutated gene does not code for a functional protein, all that is necessary is to provide the cell with the correct version of the gene: removal of the defective genes is unnecessary. The situation is less easy with dominant genetic diseases (p. 286), as with these it is the defective gene product itself that is responsible for the disease state, and so the therapy must include not only addition of the correct gene but also removal of the defective version. This requires a gene delivery system that promotes recombination between the chromosomal and vector-borne versions of the gene, so that the defective chromosomal copy is replaced by the gene from the vec-

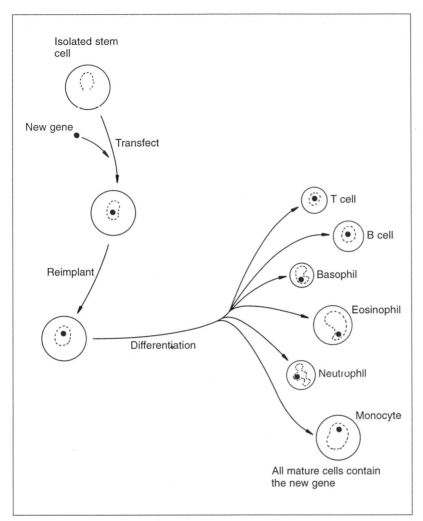

Figure 13.12 Differentiation of a transfected stem cell leads to the new gene being present in all the mature blood cells.

tor. The technique is complex and unreliable, and broadly applicable procedures have not yet been developed.

13.3.2 Ethical questions raised by gene therapy

Should gene therapy be used to cure human disease? As with many ethical questions there is no simple answer. On the one hand, there could surely be no justifiable objection to the routine application via a respiratory inhaler of correct versions of the cystic fibrosis gene as a means of managing this disease. Similarly, if bone marrow transplants are morally acceptable, then it is difficult to argue against gene therapies aimed at correction of blood disorders via stem cell transfection.

Germline therapy is a more difficult issue. The problem is that the techniques used for germline correction of inherited disease are exactly the same techniques that could be used for germline manipulation of other inherited characteristics. Indeed, the development of this technique with animals has not been prompted by any desire to cure genetic diseases, the aims being to 'improve' farm animals, for example by making genetic changes that result in lower fat content. This type of manipulation, where the genetic constitution of an organism is changed in a directed, heritable fashion, is clearly unacceptable with humans. At present, technical problems mean that human germline manipulation would be extremely difficult. Before these problems are solved we should ensure that the desire to do good should not raise the possibility of doing tremendous harm.

FURTHER READING

Goeddel, D. V. *et al.* (1979) Expression in *Escherichia coli* of chemically synthesized genes for human insulin. *Proceedings of the National Academy of Sciences, USA,* **76,** 106–10.

Itakura, K. *et al.* (1977) Expression in *Escherichia coli* of a chemically synthesized gene for the hormone somatostatin. *Science,* **198,** 1056–63.

Goeddel, D. *et al.* (1979) Direct expression in *Escherichia coli* of a DNA sequence coding for human growth hormone. *Nature,* **281,** 544–8 – production of recombinant somatotropin.

Yonemura, K. *et al.* (1993) Efficient production of recombinant human factor VIII by co-expression of the heavy and light chains. *Protein Engineering,* **6,** 669–74.

Tartaglia, J. and Paoletti, E. (1988) Recombinant vaccinia virus vaccines. *Trends in Biotechnology,* **6,** 43–6.

Valuenzuela, P. *et al.* (1985) Synthesis and assembly in yeast of hepatitis B surface antigen particles containing the polyalbumin receptor. *Biotechnology,* **3,** 317–20.

Miki, Y. *et al.* (1994) A strong candidate for the breast and ovarian cancer susceptibility gene *BRCA1. Science,* **266,** 66–71.

Black, D. M. and Solomon, E. (1993) The search for the familial breast/ovarian cancer gene. *Trends in Genetics,* **9,** 22–6.

Nowak, R. (1994) Breast cancer gene offers surprises. *Science,* **265,** 1796–9.

Weatherall, D. J. (1991) *The New Genetics and Clinical Practice,* 3rd edn, Oxford University Press, Oxford – includes a discussion of gene therapy and other issues relevant to this chapter.

Gene cloning in agriculture

14

Agriculture, or more specifically the cultivation of plants, is the world's oldest biotechnology, with an unbroken history that stretches back at least 10 000 years. Throughout this period humans have constantly searched for improved varieties of their crop plants, varieties with better nutritional qualities, higher yields, or features that aid cultivation and harvesting. During the first few millennia, crop improvements occurred in a sporadic fashion, but in recent centuries new varieties have been obtained by breeding programmes of ever-increasing sophistication. However, the most sophisticated breeding programme still retains an element of chance, dependent as it is on the random merging of parental characteristics in the hybrid offspring that are produced. The development of a new variety of crop plant, displaying a precise combination of desired characteristics, is a lengthy and difficult process.

Gene cloning provides a new dimension to crop breeding by enabling directed changes to be made to the genotype of a plant, circumventing the random processes inherent in conventional breeding. Two general strategies have been used:

1. **Gene addition**, in which cloning is used to alter the characteristics of a plant by providing it with one or more new genes.
2. **Gene subtraction**, in which genetic engineering techniques are used to inactivate one or more of the plant's existing genes.

A number of projects are being carried around the world, many by biotechnology companies, aimed at exploiting the potential of gene addition and gene subtraction in crop improvement. In this chapter we will investigate a representative selection of these projects, and look at some of the problems that must be solved if plant genetic engineering is to gain widespread acceptance in agriculture.

14.1 THE GENE ADDITION APPROACH TO PLANT GENETIC ENGINEERING

Gene addition simply involves the use of cloning techniques to introduce into a plant one or more new genes coding for a useful characteristic that the plant lacks. A good example of the technique is provided by the development of plants that resist insect attack by synthesizing insecticides coded by cloned genes.

14.1.1 Plants that make their own insecticides

Plants are subject to predation by virtually all other types of organism – viruses, bacteria, fungi and animals – but in agricultural settings the greatest problems are caused by insects. To reduce losses, crops are regularly sprayed with insecticides. Most conventional insecticides (e.g. pyrethroids and organophosphates) are relatively non-specific poisons that kill a broad spectrum of insects, not just the ones eating the crop. Because of their high toxicity, several of these insecticides also have potentially harmful side effects for other members of the local biosphere, including in some cases humans. These problems are exacerbated by the need to apply conventional insecticides to the surfaces of plants by spraying, which means that subsequent movement of the chemicals in the ecosystem cannot be controlled. Furthermore, insects that live within the plant, or on the undersurfaces of leaves, can sometimes avoid the toxic effects altogether.

What features would be displayed by the ideal insecticide? Clearly it must be toxic to the insects against which it is targeted, but if possible this toxicity should be highly selective, so that the insecticide is harmless to other insects and is not poisonous to animals and to humans. The insecticide should be biodegradable, so that any residues that remain after the crop is harvested, or which are carried out of the field by rainwater, do not persist long enough to damage the environment. And it should be possible to apply the insecticide in such a way that all parts of the crop, not just the upper surfaces of the plants, are protected against insect attack.

The ideal insecticide has not yet been discovered. The closest we have are the ∂-endotoxins produced by the soil bacterium *Bacillus thuringiensis*.

(a) The ∂-endotoxins of *Bacillus thuringiensis* Insects not only eat plants: bacteria also form an occasional part of their diet. In response, several types of bacteria have evolved defence mechanisms against insect predation, an example being *Bacillus thuringiensis* which, during sporulation, forms intracellular crystalline bodies that contain an insecticidal protein called the ∂-

endotoxin. The activated protein is highly poisonous to insects, some 80 000 times more toxic than organophosphate insecticides, and is relatively selective, different strains of the bacterium synthesizing proteins effective against the larvae of different groups of insects (Table 14.1).

Table 14.1 The range of insects poisoned by the various types of *B. thuringiensis* ∂-endotoxins

∂-Endotoxin type	Effective against
CryI	Lepidoptera (moth and butterfly) larvae
CryII	Lepidoptera and Diptera (two-winged fly) larvae
CryIII	Lepidoptera larvae
CryIV	Diptera larvae
CryV	Nematode worms
CryVI	Nematode worms

The ∂-endotoxin protein that accumulates in the bacterium is in fact an inactive precursor. After ingestion by the insect this protoxin is cleaved by proteinases, resulting in shorter versions of the protein that display the toxic activity, binding to the inside of the insect's gut and damaging the surface epithelium so that the insect is unable to feed and consequently starves to death (Figure 14.1). Variation in the structure of these binding sites in different groups of insects is probably the underlying cause of the high specificities displayed by the different types of ∂-endotoxin.

B. thuringiensis toxins are not recent discoveries, the first patent for their use in crop protection having been granted in 1904. Over the years there have been several attempts to market them as environmentally friendly insecticides, but their biodegradability acts as a disadvantage because it means that they must be reapplied at regular intervals during the growing season, increasing the farmer's costs. Current research is therefore aimed at developing ∂-endotoxins that do not require regular application. One approach is via protein engineering (p. 225), modifying the structure of the toxin so that it is more stable. A second approach is to engineer the crop to synthesize its own toxin.

(b) Cloning a ∂-endotoxin gene in maize Maize is an example of a crop plant that is not served well by conventional insecticides. A major pest is the European corn borer (*Ostrinia nubilialis*), which tunnels into the plant from eggs laid on the undersurfaces of leaves, thereby evading the effects of insecticides applied by spraying. The first attempt at countering this pest by engineering maize plants to synthesize ∂-endotoxin was made by plant biotechnolo-

Figure 14.1 Mode of action of a
∂-endotoxin.

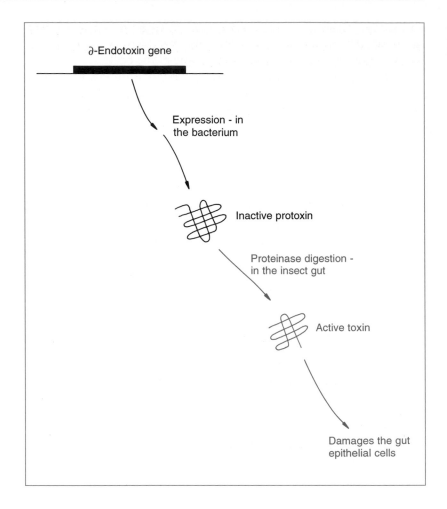

gists at a Ciba–Geigy laboratory in North Carolina. They worked
with the CrylA(b) version of the toxin, which had previously been
shown to be a 1155-amino-acid protein, with the toxic activity
residing in the segment from amino acids 29 to 607. Rather than
isolating the natural gene the Ciba–Geigy group made a shortened
version, containing the first 648 codons, by artificial gene synthesis.
This strategy enabled them to introduce modifications into the
gene to improve its expression in maize plants. For example, the
codons that were used in the artificial gene were those known to be
preferred by maize, and the overall GC content of the gene was set
at 65%, compared with the 38% GC content of the native bacterial
version of the gene (Figure 14.2(a)). The artificial gene was ligated
into a cassette vector (p. 261) between a promoter and polyadenylation signal from cauliflower mosaic virus (Figure 14.2(b)), and
introduced into maize embryos by bombardment with DNA-
coated microprojectiles (pp. 104 and 152). The embryos were

grown into mature plants, and transformants identified by PCR analysis of DNA extracts (p. 000), using primers specific for a segment of the artificial gene (Figure 14.2(c)).

The next step was to use an immunological test to determine if ∂-endotoxin was being synthesized by the transformed plants. The results showed that the artificial gene was indeed active, but that the amounts of ∂-endotoxin being produced varied from plant to

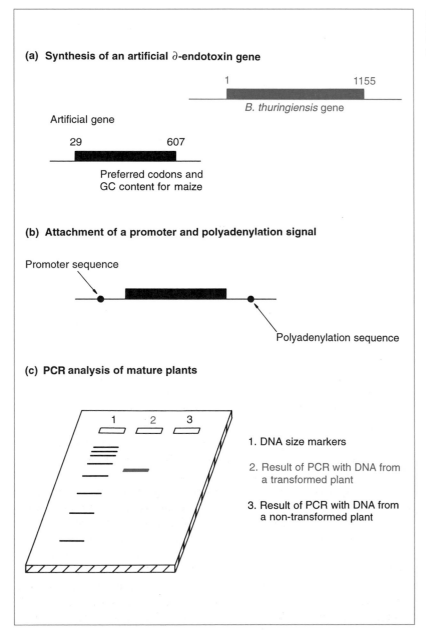

Figure 14.2 Important steps in the procedure used to obtain genetically engineered maize plants expressing an artificial ∂-endotoxin gene.

plant, from about 250 to 1750 ng of toxin per mg of total protein. These differences were probably due to **positional effects**, the level of expression of a gene cloned in a plant (or animal) often being influenced by the exact location of the gene in the host chromosomes. For reasons that are not fully understood, genes inserted at some positions are less well expressed than genes located at other positions (Figure 14.3).

Figure 14.3 Positional effects.

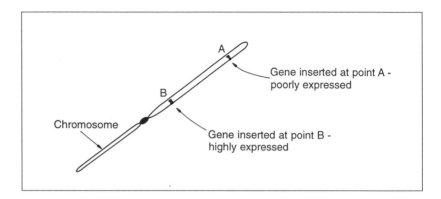

Were the transformed plants able to resist the attentions of the corn borers? This was assessed by field trials in which transformed and normal maize plants were artificially infested with larvae and the effects of predation measured over a period of 6 weeks. Two criteria were used: the amount of damage suffered by the foliage of the infested plants, and the lengths of the tunnels produced by the larvae boring into the plants. In both respects the transformed plants gave better results than the normal ones. In particular, the average lengths of the larval tunnels was reduced from 40.7 cm with the controls to just 6.3 cm for the engineered plants. In real terms, this is a very significant level of resistance.

14.1.2 Other gene addition projects

Maize is not the only plant that has been engineered to produce ∂-endotoxin, similar projects having been carried out with rice, cotton, potato, tomato and other crops. Neither is this the only approach to insect resistance. Equally successful results have been obtained with genes coding for proteinase inhibitors, small polypeptides that disrupt the activities of enzymes in the insect gut, preventing or slowing growth. Proteinase inhibitors are produced naturally by several types of plant, notably legumes such as cowpeas and common beans, and their genes have been successfully transferred to other crops which do not normally make significant amounts of these proteins. The inhibitors are particularly effective against beetle larvae that feed on seeds, and so may be a better alternative than ∂-endotoxin for plants whose seeds are stored for long periods.

Examples of other gene addition projects are listed in Table 14.2. In many cases the objective is to improve the plant's ability to withstand pests such as insects, fungi, bacteria and viruses, or to be able to resist the toxic effects of herbicides used to control weeds. Other projects are starting to explore the use of genetic engineering to improve the nutritional quality of crop plants, for example by increasing the content of essential amino acids and by changing the plant biochemistry so that more of the available nutrients can be utilized during digestion by humans or animals.

Table 14.2 Examples of gene addition projects with plants

Gene for	Source organism	Characteristic conferred on transformed plants
∂-Endotoxin	Bacillus thuringiensis	Insect resistance
Proteinase inhibitors	Various legumes	Insect resistance
Chitinase	Rice	Fungal resistance
Glucanase	Alfalfa	Fungal resistance
Ribosome-inactivating protein	Barley	Fungal resistance
Ornithine carbamoyltransferase	Pseudomonas syringae	Bacterial resistance
Virus coat proteins	Various viruses	Virus resistance
Satellite RNAs	Various viruses	Virus resistance
2'-5' oligoadenylate synthetase	Rat	Virus resistance
Mutant form of 5-enol-pyruvyl-shikimic acid 3-phosphate synthase	Salmonella typhimurium	Herbicide resistance
Phosphinothricin acetyl transferase	Streptomyces hygroscopicus	Herbicide resistance
Ribonuclease	Bacillus amyloliquefaciens	Male sterility
Methionine-rich protein	Brazil nuts	Improved S content
Thaumatin	Thaumatococcus danielli	Sweetness
Monellin	Thaumatococcus danielli	Sweetness
Stearyl-CoA desaturase	Rat	Higher monounsaturated fatty acid content

14.2 GENE SUBTRACTION

Gene subtraction is a misnomer as the modification does not involve the actual removal of a gene, merely its inactivation. There are several ways by which a single, chosen gene could be inactivated in a living plant, the most successful so far in practical terms being the use of **antisense technology**. The example we will focus on is an important one as it has resulted in one of the first genetically engineered foodstuffs to be approved for sale to the general public.

14.2.1 The principle behind antisense technology

In an antisense experiment the gene to be cloned is ligated into the vector in reverse orientation (Figure 14.4). This means that when the cloned 'gene' is transcribed the RNA that is synthesized is the reverse complement of the mRNA produced from the normal version of the gene. We refer to this reverse complement as an **antisense RNA**, sometimes abbreviated to asRNA.

An antisense RNA is able to prevent synthesis of the product of the gene it is directed against. The underlying mechanism is not altogether clear, but it almost certainly involves hybridization between the antisense and sense copies of the RNA (Figure 14.5). It is possible that the block to expression arises because the resulting double-stranded RNA molecule is rapidly degraded by cellular ribonucleases, or the explanation might be that the antisense RNA simply prevents ribosomes from attaching to the sense strand. Whatever the mechanism, synthesis of antisense RNA in a transformed plant is an effective way of carrying out gene subtraction.

14.2.2 Antisense RNA and the engineering of fruit ripening in tomato

At present, commercially-grown tomatoes and other soft fruits are usually picked before they are completely ripe, to allow time for the fruits to be transported to the marketplace before they begin to spoil. This is essential if the process is to be economically viable, but there is a problem in that most immature fruits do not develop their full flavour if they are removed from the plant before they are fully ripe. The result is that mass-produced tomatoes often have a bland taste which makes them less attractive to the consumer. Two biotechnology companies, Calgene in the USA and ICI Seeds in the UK, have used antisense technology as a means of genetically engineering tomato plants so that the fruit ripening process is slowed down. This enables the grower to leave the fruits on the plant until they ripen to the stage where the flavour has fully developed, there still being time to transport and market the crop before spoilage sets in.

(a) The role of the polygalacturonase gene in tomato fruit ripening The timescale for development of a fruit is measured as the number of days or weeks after flowering. In tomato, this

Figure 14.4 Antisense RNA.

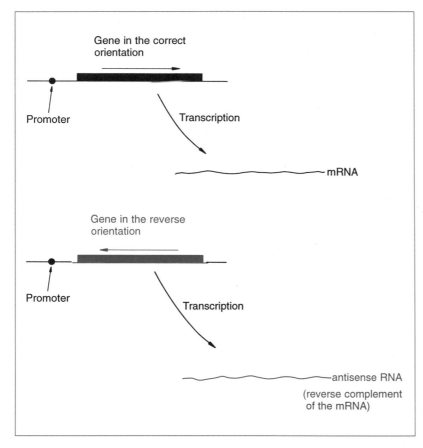

process takes approximately 8 weeks from start to finish, with the colour and flavour changes associated with ripening beginning after about 6 weeks. At about this time a number of genes involved in the later stages of ripening are switched on, including one coding for the polygalacturonase enzyme (Figure 14.6). This enzyme slowly breaks down the polygalacturonic acid component of the cell walls in the fruit pericarp, resulting in a gradual softening. The softening makes the fruit palatable, but if taken too far results in a squashy, spoilt tomato attractive only to students with limited financial resources.

Partial inactivation of the polygalacturonase gene might increase the time between flavour development and spoilage of the fruit. How could antisense technology be used to achieve this result?

(b) Cloning the antisense polygalacturonase 'gene'   The experiment that we will follow was carried out by the Plant Biotechnology Section of ICI Seeds, together with scientists at the University of Nottingham, in the mid-1980s. A 730-bp restriction fragment was obtained from the 5′ region of the normal polygalacturonase gene,

Figure 14.5 Possible mechanisms for the inhibition of gene expression by antisense RNA.

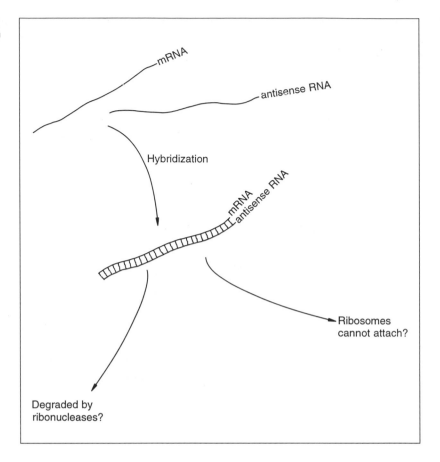

representing just under half of the coding sequence (Figure 14.7). A plant polyadenylation signal was attached to the beginning of this fragment, a cauliflower mosaic virus promoter was ligated to the end, and the construction was inserted into the Ti plasmid vector pBIN19 (p. 147). Once inside the plant, transcription from the cauliflower mosaic virus promoter should result in synthesis of an antisense RNA complementary to the first half of the polygalacturonase mRNA. Previous experiments with antisense RNA had suggested that this would be sufficient to reduce or even prevent translation of the target mRNA.

Transformation was carried out by introducing the recombinant pBIN19 molecules into *Agrobacterium tumefaciens* bacteria and then allowing the bacteria to infect tomato stem segments (Figure 14.8). Small amounts of callus material collected from the surfaces of these segments were tested for their ability to grow on an agar medium containing kanamycin (remember that pBIN19 carries a gene for kanamycin resistance – see Figure 7.14). Resistant transformants were identified and allowed to develop into mature plants.

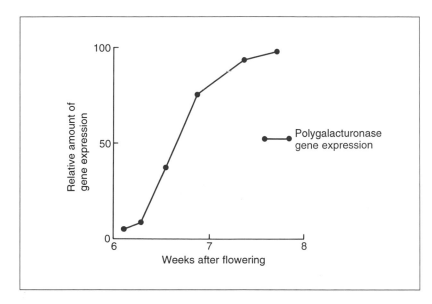

Figure 14.6 The increase in poly-galacturonase gene expression seen during the later stages of tomato fruit ripening.

The results of the experiment were assessed in a number of ways:

1. The presence of the antisense 'gene' in the DNA of the trans-formed plants was checked by Southern hybridization (p. 184).
2. Expression of the antisense 'gene' was measured by northern hybridization (p. 184) with a single-stranded DNA probe that would hybridize only to the antisense RNA.
3. The effect of antisense RNA synthesis on the amount of

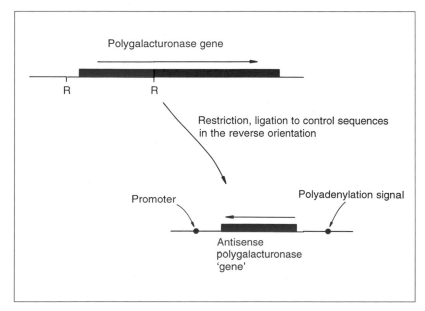

Figure 14.7 Construction of an anti-sense polygalacturonase 'gene'. R, restriction site.

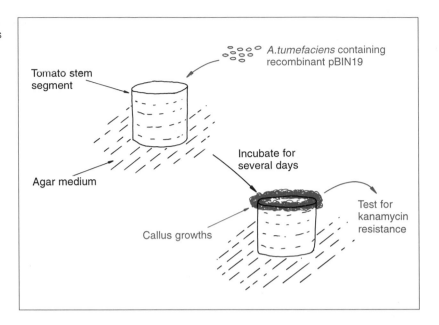

Figure 14.8 Obtaining transformed plants by infection of stem segments with recombinant *A. tumefaciens*.

polygalacturonase mRNA in the cells of ripening fruit was determined by northern hybridization with a second single-stranded DNA probe, this one specific for the sense mRNA. These experiments showed that ripening fruit from transformed plants contained less polygalacturonase mRNA than the fruits from normal plants.

4. The amounts of polygalacturonase enzyme produced in the ripening fruits of transformed plants were estimated from the intensities of the relevant bands after separation of fruit proteins by polyacrylamide gel electrophoresis, and by directly measuring the enzyme activities in the fruits. The results showed that less enzyme was synthesized in transformed fruits (Figure 14.9).

Most importantly, the transformed fruits, although undergoing a gradual softening, could be stored for a prolonged period before beginning to spoil. This indicated that the antisense RNA had not completely inactivated the polygalacturonase gene, but had nonetheless produced a sufficient reduction in gene expression to delay the ripening process as desired.

14.2.3 Other examples of the use of antisense RNA in plant genetic engineering

In general terms, the applications of gene subtraction in plant genetic engineering are probably less broad than those of gene addition. It is easier to think of useful characteristics that a plant lacks and which might be introduced by gene addition, than it is to

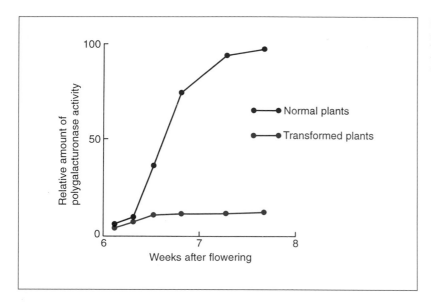

Figure 14.9 The differences in poly-galacturonase activity in normal tomato fruits and in fruits expressing the antisense polygalacturonase 'gene'.

identify disadvantageous traits that the plant already possesses and which could be removed by gene subtraction. There are, however, a growing number of plant biotechnology projects based on gene subtraction (Table 14.3), and the approach is likely to increase in importance as the uncertainties that surround the underlying principles of antisense technology are gradually resolved.

Table 14.3   Examples of gene subtraction projects with plants

Target gene	Engineered characteristic
Polygalacturonase	Delay of fruit spoilage
Polyphenol oxidase	Prevention of discolouration in fruits and vegetables
Starch synthase	Reduction of starch content
Chalcone synthase	Modification of flower colour

14.3 PROBLEMS WITH GENETICALLY ENGINEERED PLANTS

Ripening-delayed tomatoes produced by gene subtraction were the first genetically engineered whole food to be approved for marketing. Partly because of this, plant genetic engineering has provided the battleground on which biotechnologists and other interested parties have fought over the safety and ethical issues that arise from our ability to alter the genetic makeup of living organisms. A number of the most important questions do not directly concern genes and the expertise needed to answer them will not be found in this book. We cannot discuss in an authoritative fashion the possible impact, good or otherwise, that genetically engineered crops might have on local farming practices in the developing world. Neither can we comment on the motives behind the development of herbicide resistant plants by companies who also market the herbicide which farmers will have to use with the engineered crop. However, we can, and should, look at the biological issues.

14.3.1 Safety concerns with selectable markers

One of the main areas of concern to emerge from the debate over genetically engineered tomatoes is the possible harmful effects of the marker genes used with plant cloning vectors. Most plant vectors carry a copy of a gene for kanamycin resistance, enabling transformed plants to be identified during the cloning process. The *kan^R* gene, also called *nptII*, is bacterial in origin and codes for the enzyme neomycin phosphotransferase II. This gene, and its enzyme product, are present in all cells of an engineered plant. The fear that neomycin phosphotransferase might be toxic to humans has been allayed by tests with animal models, but two other safety issues remain:

1. Could the *kan^R* gene contained in a genetically engineered food-stuff be passed to bacteria in the human gut, making these resistant to kanamycin and related antibiotics?
2. Could the *kan^R* gene be passed to other organisms in the environment and would this result in damage to the ecosystem?

Neither question can be fully answered with our current knowledge. It can be argued that digestive processes would destroy all the *kan^R* genes in a genetically engineered food before they could reach the bacterial flora of the gut, and that even if a gene did avoid destruction then the chances of it being transferred to a bacterium would be very small. Nevertheless, the risk factor is not zero. Similarly, although experiments suggest that growth of genetically engineered plants would have a negligible effect on the environment, as *kan^R* genes are already common in natural ecosystems, the future occurrence of some unforeseen and damaging event cannot be considered an absolute impossibility.

The fears surrounding the use of *kan^R* and other marker genes have prompted biotechnologists to devise ways of removing these genes from plant DNA after the transformation event has been verified. One of the strategies being investigated makes use of an enzyme from bacteriophage P1, called Cre, which catalyses a recombination event that excises DNA fragments flanked by specific 34-bp recognition sequences (Figure 14.10). To use this system the plant would have to be transformed with two cloning vectors, the first carrying the gene being added to the plant along with its *kan*^R selectable marker gene, the latter surrounded by the Cre target sequences, and the second carrying the Cre gene. After transformation, expression of the Cre gene would result in excision of the *kan*^R gene from the plant DNA.

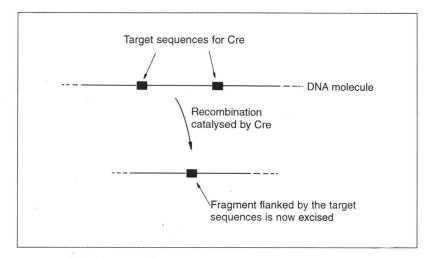

Figure 14.10 DNA excision by the Cre recombinase enzyme.

What if the Cre gene is itself hazardous in some way? This is immaterial as the two vectors used in the transformation would probably integrate their DNA fragments into different chromosomes, so random segregation during sexual reproduction would result in first generation plants that contained one integrated fragment but not the other. A plant that contains neither the Cre gene nor the *kan*^R selectable marker, but does contain the important gene that we wished to add to the plant's genome, could therefore be obtained.

14.3.2 The possibility of harmful effects on the environment

A second area of concern regarding genetically engineered plants is that their new gene combinations might harm the environment in some way. Particular problems have been highlighted with plants engineered to be resistant to virus infection. One approach here is to transform the plant with genes coding for coat proteins of a pathogenic virus. Expression of these genes does not result in disease symptoms but does provide the plant with a degree of protection against infection by the intact virus. One fear is that a plant that is synthesizing coat proteins for a pathogenic virus might be attacked by a second type of virus, the replication of which might lead to hybrid progeny containing genomes of the infecting virus packaged into coat proteins synthesized by the plant. These hybrids might have unexpected and dangerous properties, for example the new coat proteins might extend the host range of the second virus, enabling it to infect new plants that are normally resistant to the resulting disease.

Predicting the effects that genetically engineered plants might have on the environment will be possible only if suitable experiments are carried out, in model systems, before release of the plants is allowed. Regulations regarding release vary from country to country, but throughout most of Europe and North America the requirements are so stringent, and the opposition so strong and well-informed, that the potential for damage is being minimized. In the long run, however, it is the responsibility of genetic engineers to ensure that it is safe for the products of their experiments to be released into the environment and safe for these products to be consumed. Projects aimed at assessing the risks are therefore just as important as the genetic engineering experiments themselves.

FURTHER READING

Feitelson, J. S., Payne, J. and Kim, L. (1992) *Bacillus thuringiensis*: insects and beyond. *Biotechnology*, **10**, 271–5 – details of ∂-endotoxins and their potential as conventional insecticides and in genetic engineering.

Fischhoff, D. A. *et al.* (1987) Insect tolerant transgenic tomato plants. *Biotechnology*, **5**, 807–13 – the first transfer of a *B. thuringiensis* ∂-endotoxin gene to a plant.

Koziel, M. G. *et al* (1993) Field performance of elite transgenic maize plants expressing an insecticidal protein derived from *Bacillus thuringiensis. Biotechnology*, **11**, 194–200.

Shade, R. E. *et al.* (1994) Transgenic pea seeds expressing the α-amylase inhibitor of the common bean are resistant to bruchid beetles. *Biotechnology*, **12**, 793–6 – a second approach to insect-resistant plants.

Truve, E. *et al.* (1993) Transgenic potato plants expressing mammalian 2′–5′ oligoadenylate synthetase are protected from potato virus X infection under field conditions. *Biotechnology*, **11**, 1048–52 – engineering virus resistance by gene addition.

Smith, C. J. S *et al.* (1988) Antisense RNA inhibition of polygalacturonase gene expression in transgenic tomatoes. *Nature*, **334**, 724–6.

Bachem, C. W. B. *et al.* (1994) Antisense expression of polyphenol oxidase genes inhibits enzymatic browning in potato tubers. *Biotechnology*, **12**, 1101–5.

Courtney-Gutterson, N. *et al.* (1994) Modification of flower color in florist's chrysanthemum: production of a white-flowering variety through molecular genetics. *Biotechnology*, **12**, 268–71 – another example of the use of antisense RNA

Hoyle, R. (1993) Herbicide-resistant crops are no conspiracy. *Biotechnology*, **11**, 783–4.

Flavell, R. B., Dart, E., Fuchs, R. L. and Fraley, R. T. (1992) Selectable marker genes: safe for plants? *Biotechnology*, **10**, 141–4 – describes the possible hazards posed by *kan^R* and other marker genes in engineered plants.

Yoder, J. I. and Goldsbrough, A. P. (1994) Transformation systems for generating marker-free transgenic plants. *Biotechnology*, **12**, 263–7.

Tepfer, M. (1993) Viral genes and transgenic plants: what are the potential environmental risks? *Biotechnology*, **11**, 1125–32 – an excellent exploration of this difficult topic.

What to read next?

Now that you have finished this book you can move on to the more advanced texts on gene cloning. As well as descriptions of techniques and applications, these books also contain detailed procedures to follow when doing the experiments. The original of these cloning manuals, and still revered as 'The Bible', is *Molecular Cloning: A Laboratory Manual*, 2nd edn (1989) by J. Sambrook *et al.*, Cold Spring Harbor Laboratory Press, Cold Spring Harbor, New York. Equally comprehensive, and with the advantage of being updated every few months, is *Current Protocols in Molecular Biology* (first published in 1987) by F. M. Ausubel *et al.*, Greene Publishing Associates and John Wiley and Sons, New York. If you have one or both of these books then you will not go far wrong in your first attempts at practical gene cloning.

There are also a few books that attempt to explain the applications of gene cloning in research and biotechnology. The most successful of these are *The New Genetics and Clinical Practice*, 3rd edn (1991) by D. J. Weatherall, Oxford University Press, Oxford, and *Molecular Biotechnology*, 2nd edn (1991) by S. B. Primrose, Blackwell Scientific Publications, Oxford.

Gene cloning techniques and biotechnology are advancing at a great pace and scientific journals offer the only means of keeping in touch with the developments. The best general interest journal by some way is *Science*, which regularly contains informative and detailed articles on all aspects of molecular biology, including the many applications of gene cloning. Two journals devoted to the industrial aspects of gene cloning, *Biotechnology* and *Trends in Biotechnology*, are also highly recommended.

Glossary

Adaptor A synthetic, double-stranded oligonucleotide used to attach sticky ends to a blunt-ended molecule.

Adenovirus An animal virus, derivatives of which have been used to clone genes in mammalian cells.

Affinity chromatography A chromatography method that makes use of a ligand that binds a specific protein and which can therefore be used to aid purification of that protein.

Agrobacterium tumefaciens The soil bacterium which, when containing the Ti plasmid, is able to form crown galls on a number of dicotyledonous plant species.

Antisense RNA An RNA molecule that is the reverse complement of a naturally occurring mRNA, and which can be used to prevent translation of that mRNA in a transformed cell.

Antisense technology The use in genetic engineering of a gene coding for an antisense RNA.

Autoradiography A method of detecting radioactively labelled molecules through exposure of an X-ray-sensitive photographic film.

Auxotroph A mutant microorganism that will grow only if supplied with a nutrient not required by the wild-type.

Avidin A protein that has a high affinity for biotin and is used in the detection system for biotinylated probes.

Bacterial artificial chromosome A cloning vector based on the F plasmid, used for cloning relatively large fragments of DNA in *E. coli*.

Bacteriophage or **Phage** A virus whose host is a bacterium. Bacteriophage DNA molecules are often used as cloning vectors.

Baculovirus A virus that has been used as a cloning vector for the production of recombinant protein in insect cells.

Batch culture Growth of bacteria in a fixed volume of liquid medium in a closed vessel, with no additions or removals made during the period of incubation.

Biolistics   A means of introducing DNA into cells that involves bombardment with high-velocity microprojectiles coated with DNA.

Biological containment   One of the precautionary measures taken to prevent the replication of recombinant DNA molecules in microorganisms in the natural environment. Biological containment involves the use of vectors and host organisms that have been modified so that they will not survive outside of the laboratory.

Biotechnology   The use of living organisms, often but not always microbes, in industrial processes.

Biotin   A molecule that can be incorporated into dUTP and used as a non-radioactive label for a DNA probe.

Blunt end or **Flush end**   An end of a DNA molecule at which both strands terminate at the same nucleotide position with no single-stranded extension.

Broth culture   Growth of microorganisms in a liquid medium.

Buoyant density   The density possessed by a molecule or particle when suspended in an aqueous salt or sugar solution.

Capsid   The protein coat that encloses the DNA or RNA molecule of a bacteriophage or virus.

Cassette   A DNA sequence consisting of promoter – ribosome binding site – unique restriction site – terminator (or, for a eukaryotic host, promoter – unique restriction site – polyadenylation sequence) carried by certain types of expression vector. A foreign gene inserted into the unique restriction site will be placed under control of the expression signals.

Cauliflower mosaic virus (CaMV)   The best studied of the caulimoviruses, used as a cloning vector for some species of higher plant.

Caulimoviruses   One of the two groups of DNA viruses to infect plants, the members of which have potential as cloning vectors for some species of higher plant.

Cell extract   A preparation consisting of a large number of broken cells and their released contents.

Cell-free translation system   A cell extract containing all the components required for protein synthesis (i.e. ribosomal subunits, tRNAs, amino acids, enzymes and cofactors) and able to translate added mRNA molecules.

Chimaera   A recombinant DNA molecule made up of DNA fragments from more than one organism, named after the mythological beast.

Chromosome   A self-replicating nucleic acid molecule carrying a number of genes.

Chromosome walking   A technique used to identify a series of overlapping restriction fragments, often to determine the relative positions of genes on large DNA molecules.

Cleared lysate   A cell extract that has been centrifuged to remove cell debris, subcellular particles and possibly chromosomal DNA.

Clone   A population of identical cells, generally those containing identical recombinant DNA molecules.

Compatibility   Refers to the ability of two different types of plasmid to coexist in the same cell.

Competent   Refers to a culture of bacteria that have been treated to enhance their ability to take up DNA molecules.

Complementary   Refers to two polynucleotides that can base-pair to form a double-stranded molecule.

Complementary DNA (cDNA) cloning   A cloning technique involving conversion of purified mRNA to DNA before insertion into a vector.

Conformation   The spatial organization of a molecule. Linear and circular are two possible conformations of a polynucleotide.

Conjugation   Physical contact between two bacteria, usually associated with transfer of DNA from one cell to the other.

Consensus sequence   A nucleotide sequence used to describe a large number of related though non-identical sequences. Each position of the consensus sequence represents the nucleotide most often found at that position in the real sequences.

Continuous culture   The culture of microorganisms in liquid medium under controlled conditions, with additions to and removals from the medium over a lengthy period of time.

Contour clamped homogeneous electric fields (CHEF)   An electrophoresis technique for the separation of large DNA molecules.

Copy number   The number of molecules of a plasmid contained in a single cell.

***Cos* site**   One of the cohesive, single-stranded extensions present at the ends of the DNA molecules of certain strains of λ phage.

Cosmid   A cloning vector consisting of the λ *cos* site inserted into a plasmid, used to clone DNA fragments up to 40 kb in size.

Covalently closed-circular (CCC)   A completely double-stranded circular DNA molecule, with no nicks or discontinuities, usually with a supercoiled conformation.

Defined medium   A bacterial growth medium in which all the components are known.

Deletion analysis   The identification of control sequences for a gene by determining the effects on gene expression of specific deletions in the upstream region.

Denaturation   Of nucleic acid molecules: breakdown by chemical or physical means of the hydrogen bonds involved in base-pairing.

Density-gradient centrifugation   Separation of molecules and particles on the basis of buoyant density, by centrifugation in a concentrated sucrose or caesium chloride solution.

Deoxyribonuclease   An enzyme that degrades DNA.

Dideoxynucleotide   A modified nucleotide that lacks the 3′ hydroxyl group and so prevents further chain elongation when incorporated into a growing polynucleotide.

Disarmed plasmid   A Ti plasmid that has had some or all of the T-DNA genes removed, so it is no longer able to promote cancerous growth of plant cells.

DNA sequencing   Determination of the order of nucleotides in a DNA molecule.

Double digestion   Cleavage of a DNA molecule with two different restriction endonucleases, either concurrently or consecutively.

Electrophoresis   Separation of molecules on the basis of their charge-to-mass ratio.

Electroporation   A method for increasing DNA uptake by protoplasts through prior exposure to a high voltage which results in the temporary formation of small pores in the cell membrane.

End-filling   Conversion of a sticky end to a blunt end by enzymatic synthesis of the complement to the single-stranded extension.

Endonuclease   An enzyme that breaks phosphodiester bonds within a nucleic acid molecule.

Episome   A plasmid capable of integration into the host cell's chromosome.

Ethanol precipitation   Precipitation of nucleic acid molecules by ethanol plus salt, used primarily as a means of concentrating DNA.

Ethidium bromide   A fluorescent chemical that intercalates between base pairs in a double-stranded DNA molecule, used in the detection of DNA.

Exonuclease   An enzyme that sequentially removes nucleotides from the ends of a nucleic acid molecule.

Expression vector A cloning vector designed so that a foreign gene inserted into the vector will be expressed in the host organism.

Fermenter A vessel used for the large-scale culture of microorganisms.

Field inversion gel electrophoresis (FIGE) An electrophoresis technique for the separation of large DNA molecules.

Fluorescence *in situ* hybridization (FISH) A hybridization technique that uses fluorochromes of different colours to enable two or more genes to be located within a chromosome preparation in a single *in situ* experiment.

Footprinting The identification of a protein-binding site on a DNA molecule by determining which phosphodiester bonds are protected from cleavage by DNase I.

Gel electrophoresis Electrophoresis performed in a gel matrix so that molecules of similar electric charge can be separated on the basis of size.

Gel retardation A technique that identifies a DNA fragment that has a bound protein by virtue of its decreased mobility during gel electrophoresis.

Geminivirus One of the two groups of DNA viruses that infect plants, the members of which have potential as cloning vectors for some species of higher plants.

Gene A segment of DNA that codes for an RNA and/or polypeptide molecule.

Gene addition A genetic engineering strategy that involves the introduction of a new gene or group of genes into an organism.

Gene cloning Insertion of a fragment of DNA, carrying a gene, into a cloning vector, and subsequent propagation of the recombinant DNA molecule in a host organism.

Gene mapping Determination of the relative positions of different genes on a DNA molecule.

Gene subtraction A genetic engineering strategy that involves the inactivation of one or more of an organism's genes.

Gene therapy A method that attempts to cure an inherited disease by providing the patient with a functioning copy of the defective gene.

Genetic engineering The use of experimental techniques to produce DNA molecules containing new genes or new combinations of genes.

Genetic fingerprinting   A hybridization technique that detects the organization of highly polymorphic target sequences and which can be used to produce a banding pattern that is unique for each individual.

Genetics   The branch of biology devoted to the study of genes.

Genome   The complete set of genes of an organism.

Genomic library   A collection of clones sufficient in number to include all the genes of a particular organism.

Harvesting   The removal of microorganisms from a culture, usually by centrifugation.

Helper phage   A phage that is introduced into a host cell in conjunction with a related cloning vector, in order to provide enzymes required for replication of the cloning vector.

Heterologous probing   The use of a labelled nucleic acid molecule to identify related molecules by hybridization probing.

Homology   Refers to two genes, from different organisms and therefore of different sequence, that code for the same gene product. Two homologous genes are usually sufficiently similar in sequence for one to be used as a hybridization probe for the other.

Homopolymer tailing   The attachment of a sequence of identical nucleotides (e.g. AAAAA) to the end of a nucleic acid molecule, usually referring to the synthesis of single-stranded homopolymer extensions on the ends of a double-stranded DNA molecule.

Horseradish peroxidase   An enzyme that can be complexed to DNA and which is used in a non-radioactive procedure for DNA labelling.

Host-controlled restriction   A mechanism by which some bacteria prevent phage attack through the synthesis of a restriction endonuclease that cleaves the non-bacterial DNA.

Hybrid-arrest translation (HART)   A method used to identify the polypeptide coded by a cloned gene.

Hybridization probe   A labelled nucleic acid molecule that can be used to identify complementary or homologous molecules through the formation of stable base-paired hybrids.

Hybrid-release translation (HRT)   A method used to identify the polypeptide coded by a cloned gene.

Immunological screening   The use of an antibody to detect a polypeptide synthesized by a cloned gene.

Inclusion body   A crystalline or paracrystalline deposit within a cell, often containing substantial quantities of insoluble protein.

Incompatibility group   Comprises a number of different types of plasmid, often related to each other, that are unable to coexist in the same cell.

Induction   (1) Of a gene: the switching on of the expression of a gene or group of genes in response to a chemical or other stimulus. (2) Of λ phage: excision of the integrated form of λ, and switch to the lytic mode of infection, in response to a chemical or other stimulus.

Insertional inactivation   The cloning strategy whereby insertion of a new piece of DNA into a vector inactivates a gene carried by the vector.

Insertion vector   A λ vector constructed by deleting a segment of non-essential DNA.

In situ **hybridization**   A technique for gene mapping involving hybridization of a labelled sample of a cloned gene to a large DNA molecule, usually a chromosome.

In vitro **mutagenesis**   Any one of several techniques used to produce a specified mutation at a predetermined position in a DNA molecule.

In vitro **packaging**   Synthesis of infective λ particles from a preparation of λ capsid proteins and a concatamer of DNA molecules separated by *cos* sites.

Klenow fragment (of DNA polymerase I)   A DNA polymerase enzyme, obtained by chemical modification of *E. coli* DNA polymerase I, used primarily in chain termination DNA sequencing.

Labelling   The incorporation of a radioactive nucleotide into a nucleic acid molecule.

Lac selection   A means of identifying recombinant bacteria containing vectors that carry the *lacZ'* gene. The bacteria are plated on a medium that contains an analogue of lactose that gives a blue colour in the presence of β-galactosidase activity.

Lambda (λ)   A bacteriophage that infects *E. coli*, derivatives of which are extensively used as cloning vectors.

Ligase (DNA ligase)   An enzyme that repairs single-stranded discontinuities in double-stranded DNA molecules in the cell. Purified DNA ligase is used in gene cloning to join DNA molecules together.

Linkage analysis   A technique for mapping the chromosomal position of a gene by comparing its inheritance pattern with that of genes and other loci whose map positions are already known.

Linker   A synthetic, double-stranded oligonucleotide used to attach sticky ends to a blunt-ended molecule.

Lysogen A bacterium that harbours a prophage.

Lysogenic infection cycle The pattern of phage infection that involves integration of the phage DNA into the host chromosome.

Lysozyme An enzyme that weakens the cell walls of certain types of bacteria.

Lytic infection cycle The pattern of infection displayed by a phage that replicates and lyses the host cell immediately after the initial infection. Integration of the phage DNA molecule into the bacterial chromosome does not occur.

M13 A bacteriophage that infects *E. coli*, derivatives of which are extensively used as cloning vectors.

Melting temperature The temperature at which a double-stranded DNA or DNA–RNA molecule denatures.

Microinjection A method of introducing new DNA into a cell by injecting it directly into the nucleus.

Minimal medium A defined medium that provides only the minimum number of different nutrients needed for growth of a particular bacterium.

Multicopy plasmid A plasmid with a high copy number.

Multigene family A number of identical or related genes present in the same organism, usually coding for a family of related polypeptides.

Nick A single-strand break, involving the absence of one or more nucleotides, in a double-stranded DNA molecule.

Nick translation The repair of a nick with DNA polymerase I, usually to introduce labelled nucleotides into a DNA molecule.

Northern transfer A technique for transferring bands of RNA from an agarose gel to a nitrocellulose or similar membrane.

Nucleic acid hybridization Formation of a double-stranded molecule by base-pairing between complementary or homologous polynucleotides.

Oligonucleotide-directed mutagenesis An *in vitro* mutagenesis technique that involves the use of a synthetic oligonucleotide to introduce the predetermined nucleotide alteration into the gene to be mutated.

Open-circular The non-supercoiled conformation taken up by a circular double-stranded DNA molecule when one or both polynucleotides carry nicks.

Origin of replication The specific position on a DNA molecule where DNA replication begins.

Orthogonal field alternation gel electrophoresis (OFAGE)   A gel electrophoresis technique that employs a pulsed electric field to achieve separation of very large molecules of DNA.

P1 vector   A cloning vector based on the P1 bacteriophage, used for cloning relatively large fragments of DNA in *E. coli*.

Papillomaviruses   A group of mammalian viruses, derivatives of which have been used as cloning vectors.

Partial digestion   Treatment of a DNA molecule with a restriction endonuclease under such conditions that only a fraction of all the recognition sites are cleaved.

Phenotypic expression   A technique designed to maximize the transformation frequency obtained when using a plasmid vector.

Pilus   One of the structures present on the surface of a bacterium containing a conjugative plasmid, through which DNA is assumed to pass during conjugation.

Plaque   A zone of clearing on a lawn of bacteria caused by lysis of the cells by infecting phage particles.

Plasmid   A usually circular piece of DNA, primarily independent of the host chromosome, often found in bacterial and some other types of cells.

Plasmid amplification   A method involving incubation with an inhibitor of protein synthesis aimed at increasing the copy number of certain types of plasmid in a bacterial culture.

Polyethylene glycol   A polymeric compound used to precipitate macromolecules and molecular aggregates.

Polylinker   A synthetic double-stranded oligonucleotide carrying a number of restriction sites.

Polymerase chain reaction   A technique that enables multiple copies of a DNA molecule to be generated by enzymatic amplification of a target DNA sequence.

Positional effect   Refers to the variations in expression levels observed for genes inserted at different positions in a genome.

Primer   A short single-stranded oligonucleotide which, when attached by base-pairing to a single-stranded template molecule, acts as the start point for complementary strand synthesis directed by a DNA polymerase enzyme.

Promoter   The nucleotide sequence, upstream of a gene, that acts as a signal for RNA polymerase binding.

Prophage   The integrated form of the DNA molecule of a lysogenic phage.

Protease   An enzyme that degrades protein.

Protein A   A protein from the bacterium *Staphylococcus aureus* that binds specifically to immunoglobulin G (i.e. antibody) molecules.

Protein engineering   A collection of techniques, including but not exclusively gene mutagenesis, that result in directed alterations being made to protein molecules, often to improve the properties of enzymes used in industrial processes.

Protoplast   A cell from which the cell wall has been completely removed.

Radioactive marker   A radioactive atom used in the detection of a larger molecule in which it is incorporated.

Random amplified polymorphic DNA (RAPD) analysis   A PCR technique that uses short random primers to amplify fragments that are representative of the genome being studied and which can be used to make comparisons between the structures of the genomes of different organisms.

Random priming   A method for DNA labelling that utilizes random DNA hexamers which anneal to single-stranded DNA and act as primers for complementary strand synthesis by a suitable enzyme.

Recombinant   A transformed cell that contains a recombinant DNA molecule.

Recombinant DNA molecule   A DNA molecule created in the test-tube by ligating together pieces of DNA that are not normally contiguous.

Recombinant DNA technology   All the techniques involved in the construction, study and use of recombinant DNA molecules.

Recombinant protein   A polypeptide that is synthesized in a recombinant cell as the result of expression of a cloned gene.

Recombination   The exchange of DNA sequences between different molecules, occurring either naturally or as a result of DNA manipulation.

Relaxed   Refers to the non-supercoiled conformation of open-circular DNA.

Replacement vector   A λ vector designed so that insertion of new DNA is by replacement of part of the non-essential region of the λ DNA molecule.

Replica plating   A technique whereby the colonies on an agar plate are transferred *en masse* to a new plate, on which the colonies grow in the same relative positions as before.

Replicative form of M13   The double-stranded form of the M13 DNA molecule found within infected *E. coli* cells.

Reporter gene   A gene whose phenotype can be assayed in a transformed organism, and which is used in, for example, deletion analyses of regulatory regions.

Repression   The switching off of expression of a gene or a group of genes in response to a chemical or other stimulus.

Restriction analysis   Determination of the number and sizes of the DNA fragments produced when a particular DNA molecule is cut with a particular restriction endonuclease.

Restriction endonuclease   An endonuclease that cuts DNA molecules only at a limited number of specific nucleotide sequences.

Restriction fragment length polymorphism (RFLP)   A mutation that results in a detectable change in the pattern of fragments obtained when a DNA molecule is cut with a restriction endonuclease.

Restriction map   A map showing the positions of different restriction sites in a DNA molecule.

Retrovirus   A virus with an RNA genome, able to insert into a host chromosome, derivatives of which have been used to clone genes in mammalian cells.

Reverse transcriptase   An RNA-dependent DNA polymerase, able to synthesize a complementary DNA molecule on a template of single-stranded RNA.

RFLP linkage analysis   A technique that uses a closely linked RFLP as a marker for the presence of a particular allele in a DNA sample, often as a means of screening individuals for defective genes responsible for genetic diseases.

Ribonuclease   An enzyme that degrades RNA.

Ribosome binding site   The short nucleotide sequence upstream of a gene, which after transcription forms the site on the mRNA molecule to which the ribosome binds.

Ri plasmid   An *Agrobacterium rhizogenes* plasmid, similar to the Ti plasmid, used to clone genes in higher plants.

RT–PCR   A PCR technique in which the starting material is RNA. The first step in the procedure is conversion of the RNA to cDNA with reverse transcriptase.

Selectable marker   A gene carried by a vector and conferring a recognizable characteristic on a cell containing the vector or a recombinant DNA molecule derived from it.

Selection   A means of obtaining a clone containing a desired recombinant DNA molecule.

Shotgun cloning   A cloning strategy that involves the insertion of random fragments of a large DNA molecule into a vector, resulting in a large number of different recombinant DNA molecules.

Shuttle vector   A vector that can replicate in the cells of more than one organism (e.g. in *E. coli* and in yeast).

Simian virus 40 (SV40)   A mammalian virus used as the basis for a series of cloning vectors.

Southern transfer   A technique for transferring bands of DNA from an agarose gel to a nitrocellulose or similar membrane.

Sphaeroplast   A cell with a partially degraded cell wall.

Stem–loop   A hairpin structure, consisting of a base-paired stem and a non-base-paired loop, that may form in a polynucleotide.

Sticky end   An end of a double-stranded DNA molecule where there is a single-stranded extension.

Strong promoter   An efficient promoter that can direct synthesis of RNA transcripts at a relatively fast rate.

Stuffer fragment   The part of a λ replacement vector that is removed during insertion of new DNA.

Supercoiled   The conformation of a covalently closed-circular DNA molecule, which is coiled by torsional strain into the shape taken by a wound-up elastic band.

2 μm circle   A plasmid found in the yeast *Saccharomyces cerevisiae* and used as the basis for a series of cloning vectors.

T-DNA   The portion of the Ti plasmid transferred to the plant DNA.

Temperature-sensitive mutation   A mutation that results in a gene product that is functional within a certain temperature range (e.g. at less than 30°C), but non-functional at different temperatures (e.g. above 30°C).

Template   A single-stranded polynucleotide (or region of a polynucleotide) able to direct synthesis of a complementary polynucleotide.

Terminator   The short nucleotide sequence downstream of a gene that acts as a signal for termination of transcription.

5′-terminus   One of the two ends of a polynucleotide; that which carries the phosphate group attached to the 5′ position of the sugar.

3′-terminus   One of the two ends of a polynucleotide; that which carries the hydroxyl group attached to the 3′ position of the sugar.

Ti plasmid   The large plasmid found in those *Agrobacterium tumefaciens* cells able to direct crown gall formation on certain species of plants.

T_m The melting temperature of a double-stranded DNA or DNA–RNA molecule.

Total cell DNA   Consists of all the DNA present in a single cell or group of cells.

Transcript analysis   Experiment aimed at determining which portions of a DNA molecule are transcribed into RNA.

Transfection   The introduction of purified phage DNA molecules into a bacterial cell.

Transformation   The introduction of any DNA molecule into any living cell.

Transformation frequency   A measure of the proportion of cells in a population that are transformed in a single experiment.

Undefined medium   A growth medium in which not all the components have been identified.

UV absorbance spectroscopy   A method for measuring the concentration of a compound by determining the amount of ultraviolet radiation absorbed by a sample.

Vector   A DNA molecule, capable of replication in a host organism, into which a gene is inserted to construct a recombinant DNA molecule.

Vehicle   Often used as a substitute for the word 'vector', emphasizing that the vector transports the inserted gene through the cloning experiment.

Watson–Crick rules   The base-pairing rules that underlie gene structure and expression. A pairs with T, G with C.

Weak promoter   An inefficient promoter that directs synthesis of RNA transcripts at a relatively low rate.

Western transfer   A technique for transferring bands of protein from an electrophoresis gel to a membrane support.

Yeast artificial chromosome (YAC)   A cloning vector comprising the structural components of a yeast chromosome and able to clone very large pieces of DNA.

Yeast episomal plasmid (YEp)   A yeast vector carrying the 2 μm circle origin of replication.

Yeast integrative plasmid (YIp)   A yeast vector that relies on integration into the host chromosome for replication.

Yeast replicative plasmid (YRp)   A yeast vector that carries a chromosomal origin of replication.

Index

ABBREVIATIONS after an entry: g = glossary, f = figure, t = table